P9-CKE-592

Praise for *The Master Algorithm*

"With terms like 'machine learning' and 'Big Data' regularly making headlines, there is no shortage of hype-filled business books on the subject. There are also textbooks that are too technical to be accessible. For those in the middle—from executives to college students—this is the ideal book, showing how and why things really work without the heavy math. Unlike other books that proclaim a bright future, this one actually gives you what you need to understand the changes that are coming."

> —Peter Norvig, director of research, Google,
> and coauthor of *Artificial Intelligence*

"Starting with the audacious claim that all knowledge can be derived from data by a single 'master algorithm,' Domingos takes the reader on a fast-paced journey through the brave new world of machine learning. Writing breezily but with deep authority, Domingos is the perfect tour guide from whom you will learn everything you need to know about this exciting field, and a surprising amount about science and philosophy as well."

> —Duncan Watts, principal researcher, Microsoft Research,
> and author of *Six Degrees* and *Everything Is Obvious*

"[*The Master Algorithm*] does a good job of examining the field's five main techniques. . . . The subject is meaty and the author . . . has a knack for introducing concepts at the right moment."

> —*Economist*

"Domingos is a genial and amusing guide, who sneaks us around the backstage areas of the science in order to witness the sometimes personal (and occasionally acrimonious) tenor of research on the subject in recent decades. . . . This is a highly inclusive book, aimed at a wide range of readers from the merely curious to those who might be interested in pursuing a career in the field. Descriptions and discussions are presented with a commendable lack of jargon and the examples are clear and accessible."

> —*Times Higher Education*

"[An] interesting work."

—*CHOICE Reviews*

"An exhilarating venture into groundbreaking computer science."

—*Booklist*, starred review

"[An] enthusiastic but not dumbed-down introduction to machine learning. . . . [L]ucid and consistently informative. . . . With wit, vision, and scholarship, Domingos describes how these scientists are creating programs that allow a computer to teach itself. Readers . . . will discover fascinating insights."

—*Kirkus Reviews*

"[*The Master Algorithm*] opens the doorway to a world many of us never see or think about, though it has a tremendous impact on our daily lives."

—Shelf Awareness for Readers

"This book is a must-have to learn machine learning without equations. It will help you get the big picture of the several learning paradigms. Finally, the provocative idea is not only intriguing, but also very well argued."

—Data Mining Research

"If you are interested in a crash course on the enigmatic field of machine learning and the challenges for AI practitioners that lie ahead, this book is a great read."

—TechCast Global

"This book is a sheer pleasure, mixed with education. I am recommending it to all my students, those who studied machine learning, those who are about to do it, and those who are about to teach it. The author succeeds not only in presenting an accurate and entertaining journey through the methodological ideas behind machine learning, but also in embedding those ideas in a colorful tapestry of philosophical questions concerning the ultimate capacity of man to emulate itself. A must-read for both realists and futurists."

—Judea Pearl, professor of computer science, UCLA,
and winner of the A. M. Turing Award

"Machine learning is the single most transformative technology that will shape our lives over the next fifteen years. This book is a must-read—a bold and beautifully written new framework for looking into the future."

—Geoffrey Moore, author of *Crossing the Chasm*

"Machine learning is a fascinating world never before glimpsed by outsiders. Pedro Domingos initiates you to the mysterious languages spoken by its five tribes, and invites you to join in his plan to unite them, creating the most powerful technology our civilization has ever seen."

—Sebastian Seung, professor, Princeton,
and author of *Connectome*

"A delightful book by one of the leading experts in the field. If you wonder how AI will change your life, read this book."

—Sebastian Thrun, research professor, Stanford,
Google Fellow, and inventor of the Self-Driving Car

"This is an incredibly important and useful book. Machine learning is already critical to your life and work, and will only become more so. Finally, Pedro Domingos has written about it in a clear and understandable fashion."

—Thomas H. Davenport, distinguished professor,
Babson College, and author of *Competing on
Analytics* and *Big Data @ Work*

"Machine learning, known in commercial use as predictive analytics, is changing the world. This riveting, far-reaching, and inspiring book introduces the deep scientific concepts to even non-technical readers, and yet also satisfies experts with a fresh, profound perspective that reveals the most promising research directions. It's a rare gem indeed."

—Eric Siegel, founder, Predictive Analytics World,
and author of *Predictive Analytics*

The Master Algorithm

*How the Quest for the Ultimate Learning
Machine Will Remake Our World*

Pedro Domingos

BASIC BOOKS
A Member of the Perseus Books Group
New York

Copyright © 2015 by Pedro Domingos

Hardcover first published in 2015 by Basic Books,
an imprint of Perseus Books, LLC, a subsidiary of Hachette Book Group, Inc.
The Basic Books name and logo is a trademark of the Hachette Book Group.

First paperback edition: February 2018

Basic Books
Hachette Book Group
1290 Avenue of the Americas, New York, NY 10104
www.basicbooks.com

Hachette Book Group supports the right to free expression and the value of copyright. The purpose of copyright is to encourage writers and artists to produce the creative works that enrich our culture.

The scanning, uploading, and distribution of this book without permission is a theft of the author's intellectual property. If you would like permission to use material from the book (other than for review purposes), please contact permissions@hbgusa.com. Thank you for your support of the author's rights.

Printed in the United States of America

The Hachette Speakers Bureau provides a wide range of authors for speaking events. To find out more, go to www.hachettespeakersbureau.com or call (866) 376-6591.
The publisher is not responsible for websites (or their content) that are not owned by the publisher.

The Library of Congress has cataloged the hardcover as follows:
Domingos, Pedro.
    The master algorithm : how the quest for the ultimate learning machine will remake our world / Pedro Domingos.
        pages cm
    Includes index.
    ISBN 978-0-465-06570-7 (hardcover)—ISBN 978-0-465-06192-1 (ebook) 1. Knowledge representation (Information theory) 2. Artificial intelligence—Social aspects. 3. Artificial intelligence—Philosophy. 4. Cognitive science—Mathematics. 5. Algorithms. I. Title.

Q387.D66 2015
003'54—dc23

ISBN 978-0-465-09427-1 (paperback)   2015007615

LSC-C

10 9 8 7 6 5 4

TO THE MEMORY OF MY SISTER RITA,
WHO LOST HER BATTLE WITH CANCER WHILE
I WAS WRITING THIS BOOK

The grand aim of science is to cover the greatest number of experimental facts by logical deduction from the smallest number of hypotheses or axioms.

—*Albert Einstein*

Civilization advances by extending the number of important operations we can perform without thinking about them.

—*Alfred North Whitehead*

Contents

Prologue

You may not know it, but machine learning is all around you. When you type a query into a search engine, it's how the engine figures out which results to show you (and which ads, as well). When you read your e-mail, you don't see most of the spam, because machine learning filtered it out. Go to Amazon.com to buy a book or Netflix to watch a video, and a machine-learning system helpfully recommends some you might like. Facebook uses machine learning to decide which updates to show you, and Twitter does the same for tweets. Whenever you use a computer, chances are machine learning is involved somewhere.

Traditionally, the only way to get a computer to do something—from adding two numbers to flying an airplane—was to write down an algorithm explaining how, in painstaking detail. But machine-learning algorithms, also known as learners, are different: they figure it out on their own, by making inferences from data. And the more data they have, the better they get. Now we don't have to program computers; they program themselves.

It's not just in cyberspace, either: your whole day, from the moment you wake up to the moment you fall asleep, is suffused with machine learning.

Your clock radio goes off at 7:00 a.m. It's playing a song you haven't heard before, but you really like it. Courtesy of Pandora, it's been learning your tastes in music, like your own personal radio jock. Perhaps the song itself was produced with the help of machine learning. You eat breakfast and read the morning paper. It came off the printing press a few hours earlier, the printing process carefully adjusted to avoid streaking using a learning algorithm. The temperature in your house is just right, and your electricity bill noticeably down, since you installed a Nest learning thermostat.

As you drive to work, your car continually adjusts fuel injection and exhaust recirculation to get the best gas mileage. You use Inrix, a traffic prediction system, to shorten your rush-hour commute, not to mention lowering your stress level. At work, machine learning helps you combat information overload. You use a data cube to summarize masses of data, look at it from every angle, and drill down on the most important bits. You have a decision to make: Will layout A or B bring more business to your website? A web-learning system tries both out and reports back. You need to check out a potential supplier's website, but it's in a foreign language. No problem: Google automatically translates it for you. Your e-mail conveniently sorts itself into folders, leaving only the most important messages in the inbox. Your word processor checks your grammar and spelling. You find a flight for an upcoming trip, but hold off on buying the ticket because Bing Travel predicts its price will go down soon. Without realizing it, you accomplish a lot more, hour by hour, than you would without the help of machine learning.

During a break you check on your mutual funds. Most of them use learning algorithms to help pick stocks, and one of them is completely run by a learning system. At lunchtime you walk down the street, smart phone in hand, looking for a place to eat. Yelp's learning system helps you find it. Your cell phone is chock-full of learning algorithms. They're hard at work correcting your typos, understanding your spoken commands, reducing transmission errors, recognizing bar codes, and much else. Your phone can even anticipate what you're going to do next and advise you accordingly. For example, as you're finishing lunch, it discreetly alerts you that your afternoon meeting with an out-of-town visitor will have to start late because her flight has been delayed.

Night has fallen by the time you get off work. Machine learning helps keep you safe as you walk to your car, monitoring the video feed from the surveillance camera in the parking lot and alerting off-site security staff if it detects suspicious activity. On your way home, you stop at the supermarket, where you walk down aisles that were laid out with the help of learning algorithms: which goods to stock, which end-of-aisle displays to set up, whether to put the salsa in the sauce section or next to the tortilla chips. You pay with a credit card. A learning algorithm decided to send you the offer for that card and approved your application. Another one continually looks for suspicious transactions and alerts you if it thinks your card number was stolen. A third one tries to estimate how happy you are with this card. If you're a good customer but seem dissatisfied, you get a sweetened offer before you switch to another one.

You get home and walk to the mailbox. You have a letter from a friend, routed to you by a learning algorithm that can read handwritten addresses. There's also the usual junk, selected for you by other learning algorithms (oh, well). You stop for a moment to take in the cool night air. Crime in your city is noticeably down since the police started using statistical learning to predict where crimes are most likely to occur and concentrating beat officers there. You eat dinner with your family. The mayor is in the news. You voted for him because he personally called you on election day, after a learning algorithm pinpointed you as a key undecided voter. After dinner, you watch the ball game. Both teams selected their players with the help of statistical learning. Or perhaps you play games on your Xbox with your kids, and Kinect's learning algorithm figures out where you are and what you're doing. Before going to sleep, you take your medicine, which was designed and tested with the help of yet more learning algorithms. Your doctor, too, may have used machine learning to help diagnose you, from interpreting X-rays to figuring out an unusual set of symptoms.

Machine learning plays a part in every stage of your life. If you studied online for the SAT college admission exam, a learning algorithm graded your practice essays. And if you applied to business school and took the GMAT exam recently, one of your essay graders was a learning system. Perhaps when you applied for your job, a learning algorithm picked your résumé from the virtual pile and told your prospective employer: here's a strong candidate; take a look. Your latest raise may have come courtesy of another learning algorithm. If you're looking to buy a house, Zillow.com will estimate what each one you're considering is worth. When you've settled on one, you apply for a home loan, and a learning algorithm studies your application and recommends accepting it (or not). Perhaps most important, if you've used an online dating service, machine learning may even have helped you find the love of your life.

Society is changing, one learning algorithm at a time. Machine learning is remaking science, technology, business, politics, and war. Satellites, DNA sequencers, and particle accelerators probe nature in ever-finer detail, and learning algorithms turn the torrents of data into new scientific knowledge. Companies know their customers like never before. The candidate with the best voter models wins, like Obama against Romney. Unmanned vehicles pilot themselves across land, sea, and air. No one programmed your tastes into the Amazon recommendation system; a learning algorithm figured them out on its own, by generalizing from your past purchases. Google's self-driving car taught itself how to stay on the road; no engineer wrote an algorithm instructing it, step-by-step, how to get from A to B. No one knows how to program a car to drive, and no one needs to, because a car equipped with a learning algorithm picks it up by observing what the driver does.

Machine learning is something new under the sun: a technology that builds itself. Ever since our remote ancestors started sharpening stones into tools, humans have been designing artifacts, whether they're hand built or mass produced. But learning algorithms are artifacts that design other artifacts. "Computers are useless," said Picasso. "They can only give you answers." Computers aren't supposed to be creative; they're supposed to do what you tell them to. If what you tell them to do is be creative, you get machine learning. A learning algorithm is like a master craftsman: every one of its productions is different and exquisitely tailored to the customer's needs. But instead of turning stone into masonry or gold into jewelry, learners turn data into algorithms. And the more data they have, the more intricate the algorithms can be.

Homo sapiens is the species that adapts the world to itself instead of adapting itself to the world. Machine learning is the newest chapter in this million-year saga: with it, the world senses what you want and changes accordingly, without you having to lift a finger. Like a magic forest, your surroundings—virtual today, physical tomorrow—rearrange themselves as you move through them. The path you picked out between the trees and bushes grows into a road. Signs pointing the way spring up in the places where you got lost.

These seemingly magical technologies work because, at its core, machine learning is about prediction: predicting what we want, the results of our actions, how to achieve our goals, how the world will change. Once upon a time we relied on shamans and soothsayers for this, but they were much too fallible. Science's predictions are more trustworthy, but they are limited to what we can systematically observe and tractably model. Big data and machine learning greatly expand that scope. Some everyday things can be predicted by the unaided mind, from catching a ball to carrying on a conversation. Some things, try as we might, are just unpredictable. For the vast middle ground between the two, there's machine learning.

Paradoxically, even as they open new windows on nature and human behavior, learning algorithms themselves have remained shrouded in mystery. Hardly a day goes by without a story in the media involving machine learning, whether it's Apple's launch of the Siri personal assistant, IBM's Watson beating the human *Jeopardy!* champion, Target finding out a teenager is pregnant before her parents do, or the NSA looking for dots to connect. But in each case the learning algorithm driving the story is a black box. Even books on big data skirt around what really happens when the computer swallows all those terabytes and magically comes up with new insights. At best, we're left with the impression that learning algorithms just find correlations between pairs of events, such as googling "flu medicine" and having the flu. But finding correlations is to machine learning no more than bricks are to houses, and people don't live in bricks.

When a new technology is as pervasive and game changing as machine learning, it's not wise to let it remain a black box. Opacity opens the door to error and misuse. Amazon's algorithm, more than any one person, determines what books are read in the world today. The NSA's algorithms decide whether you're a potential terrorist. Climate models decide what's a safe level of carbon dioxide in the atmosphere. Stock-picking models drive the economy more than most of us do. You can't control what you don't understand, and that's why you need to understand machine learning—as a citizen, a professional, and a human being engaged in the pursuit of happiness.

This book's first goal is to let you in on the secrets of machine learning. Only engineers and mechanics need to know how a car's engine works, but every driver needs to know that turning the steering wheel changes the car's direction and stepping on the brake brings it to a stop. Few people today know what the corresponding elements of a learner even are, let alone how to use them. The psychologist Don Norman coined the term *conceptual model* to refer to the rough knowledge of a technology we need to have in order to use it effectively. This book provides you with a conceptual model of machine learning.

Not all learning algorithms work the same, and the differences have consequences. Take Amazon's and Netflix's recommenders, for example. If each were guiding you through a physical bookstore, trying to determine what's "right for you," Amazon would be more likely to walk you over to shelves you've frequented previously; Netflix would take you to unfamiliar and seemingly odd sections of the store but lead you to stuff you'd end up loving. In this book we'll see the different kinds of algorithms that companies like Amazon and Netflix use. Netflix's algorithm has a deeper (even if still quite limited) understanding of your tastes than Amazon's, but ironically that doesn't mean Amazon would be better off using it. Netflix's business model depends on driving demand into the long tail of obscure movies and TV shows, which cost it little, and away from the blockbusters, which your subscription isn't enough to pay for. Amazon has no such problem; although it's well placed to take advantage of the long tail, it's equally happy to sell you more expensive popular items, which also simplify its logistics. And we, as customers, are more willing to take a chance on an odd item if we have a subscription than if we have to pay for it separately.

Hundreds of new learning algorithms are invented every year, but they're all based on the same few basic ideas. These are what this book is about, and they're all you really need to know to understand how machine learning is changing the world. Far from esoteric, and quite aside even from their use in computers, they are answers to questions that matter to all of us: How do we learn? Is there a better way? What can we predict? Can we trust what we've learned? Rival schools of thought within machine learning have very different answers to these questions. The main ones are five in number, and we'll devote a chapter to each. Symbolists view learning as the inverse of deduction and take ideas from philosophy, psychology, and logic. Connectionists reverse engineer the brain and are inspired by neuroscience and physics. Evolutionaries simulate evolution on the computer and draw on genetics and evolutionary biology. Bayesians believe learning is a form of probabilistic inference and have their roots in statistics. Analogizers learn by extrapolating from similarity judgments and are influenced by psychology and mathematical optimization. Driven by the goal of building learning machines, we'll tour a good chunk of the intellectual history of the last hundred years and see it in a new light.

Each of the five tribes of machine learning has its own master algorithm, a general-purpose learner that you can in principle use to discover knowledge from data in any domain. The symbolists' master algorithm is inverse deduction, the connectionists' is backpropagation, the evolutionaries' is genetic programming, the Bayesians' is Bayesian inference, and the analogizers' is the support vector machine. In practice, however, each of these algorithms is good for some things but not others. What we really want is a single algorithm combining the key features of all of them: the ultimate master algorithm. For some this is an unattainable dream, but for many of us in machine learning, it's what puts a twinkle in our eye and keeps us working late into the night.

If it exists, the Master Algorithm can derive all knowledge in the world—past, present, and future—from data. Inventing it would be one of the greatest advances in the history of science. It would speed up the progress of knowledge across the board, and change the world in ways that we can barely begin to imagine. The Master Algorithm is to machine learning what the Standard Model is to particle physics or the Central Dogma to molecular biology: a unified theory that makes sense of everything we know to date, and lays the foundation for decades or centuries of future progress. The Master Algorithm is our gateway to solving some of the hardest problems we face, from building domestic robots to curing cancer.

Take cancer. Curing it is hard because cancer is not one disease, but many. Tumors can be triggered by a dizzying array of causes, and they mutate as they metastasize. The surest way to kill a tumor is to sequence its genome, figure out which drugs will work against it—without harming you, given *your* genome and medical history—and perhaps even design a new drug specifically for your case. No doctor can master all the knowledge required for this. Sounds like a perfect job for machine learning: in effect, it's a more complicated and challenging version of the searches that Amazon and Netflix do every day, except it's looking for the right treatment for you instead of the right book or movie. Unfortunately, while today's learning algorithms can diagnose many diseases with superhuman accuracy, curing cancer is well beyond their ken. If we succeed in our quest for the Master Algorithm, it will no longer be.

The second goal of this book is thus to enable *you* to invent the Master Algorithm. You'd think this would require heavy-duty mathematics and severe theoretical work. On the contrary, what it requires is stepping back from the mathematical arcana to see the overarching pattern of learning phenomena; and for this the layman, approaching the forest from a distance, is in some ways better placed than the specialist, already deeply immersed in the study of particular trees. Once we have the conceptual solution, we can fill in the mathematical details; but that is not for this book, and not the most important part. Thus, as we visit each tribe, our goal is to gather its piece of the puzzle and understand where it fits, mindful that none of the blind men can see the whole elephant. In particular, we'll see what each tribe can contribute to curing cancer, and also what it's missing. Then, step-by-step, we'll assemble all the pieces into the solution—or rather, *a* solution that is not yet the Master Algorithm, but is the closest anyone has come, and hopefully makes a good launch pad for your imagination. And we'll preview the use of this algorithm as a weapon in the fight against cancer. As you read the book, feel free to skim or skip any parts you find troublesome; it's the big picture that matters, and you'll probably get more out of those parts if you revisit them after the puzzle is assembled.

I've been a machine-learning researcher for more than twenty years. My interest in it was sparked by a book with an odd title I saw in a bookstore when I was a senior in college: *Artificial Intelligence*. It had only a short chapter on machine learning, but on reading it, I immediately became convinced that learning was the key to solving AI and that the state of the art was so primitive that maybe I could contribute something. Shelving plans for an MBA, I entered the PhD program at the University of California, Irvine. Machine learning was then a small, obscure field, and UCI had one of the few sizable research groups anywhere. Some of my classmates dropped out because they didn't see much of a future in it, but I persisted. To me nothing could have more impact than teaching computers to learn: if we could do that, we would get a leg up on every other problem. By the time I graduated five years later, the data-mining explosion was under way, and so was my path to this book. My doctoral dissertation unified symbolic and analogical learning. I've spent much of the last ten years unifying symbolism and Bayesianism, and more recently those two with connectionism. It's time to go the next step and attempt a synthesis of all five paradigms.

I had a number of different but overlapping audiences in mind when writing this book.

If you're curious what all the hubbub surrounding big data and machine learning is about and suspect that there's something deeper going on than what you see in the papers, you're right! This book is your guide to the revolution.

If your main interest is in the business uses of machine learning, this book can help you in at least six ways: to become a savvier consumer of analytics; to make the most of your data scientists; to avoid the pitfalls that kill so many data-mining projects; to discover what you can automate without the expense of hand-coded software; to reduce the rigidity of your information systems; and to anticipate some of the new technology that's coming your way. I've seen too much time and money wasted trying to solve a problem with the wrong learning algorithm, or misinterpreting what the algorithm said. It doesn't take much to avoid these fiascoes. In fact, all it takes is to read this book.

If you're a citizen or policy maker concerned with the social and political issues raised by big data and machine learning, this book will give you a primer on the technology—what it is, where it's taking us, what it does and doesn't make possible—without boring you with all the ins and outs. From privacy to the future of work and the ethics of roboticized warfare, we'll see where the real issues are and how to think about them.

If you're a scientist or engineer, machine learning is a powerful armory that you don't want to be without. The old, tried-and-true statistical tools don't get you far in the age of big (or even medium) data. You need machine learning's nonlinear chops to accurately model most phenomena, and it brings with it a new scientific worldview. The expression *paradigm shift* is used too casually these days, but I believe it's not an exaggeration to say that that's what this book describes.

If you're a machine-learning expert, you're already familiar with much of what the book covers, but you'll also find in it many fresh ideas, historical nuggets, and useful examples and analogies. Most of all, I hope the book will provide a new perspective on machine learning and maybe even start you thinking in new directions. Low-hanging fruit is all around us, and it behooves us to pick it, but we also shouldn't lose sight of the bigger rewards that lie just beyond. (Apropos of which,

I hope you'll forgive my poetic license in using the term *master algorithm* to refer to a general-purpose learner.)

If you're a student of any age—a high schooler wondering what to major in, a college undergraduate deciding whether to go into research, or a seasoned professional considering a career change—my hope is that this book will spark in you an interest in this fascinating field. The world has a dire shortage of machine-learning experts, and if you decide to join us, you can look forward to not only exciting times and material rewards but also a unique opportunity to serve society. And if you're already studying machine learning, I hope the book will help you get the lay of the land; if in your travels you chance upon the Master Algorithm, that alone makes it worth writing.

Last but not least, if you have an appetite for wonder, machine learning is an intellectual feast, and you're invited—RSVP!

CHAPTER ONE

The Machine-Learning Revolution

We live in the age of algorithms. Only a generation or two ago, mentioning the word *algorithm* would have drawn a blank from most people. Today, algorithms are in every nook and cranny of civilization. They are woven into the fabric of everyday life. They're not just in your cell phone or your laptop but in your car, your house, your appliances, and your toys. Your bank is a gigantic tangle of algorithms, with humans turning the knobs here and there. Algorithms schedule flights and then fly the airplanes. Algorithms run factories, trade and route goods, cash the proceeds, and keep records. If every algorithm suddenly stopped working, it would be the end of the world as we know it.

An algorithm is a sequence of instructions telling a computer what to do. Computers are made of billions of tiny switches called transistors, and algorithms turn those switches on and off billions of times per second. The simplest algorithm is: flip a switch. The state of one transistor is one bit of information: one if the transistor is on, and zero if it's off. One bit somewhere in your bank's computers says whether your account is overdrawn or not. Another bit somewhere in the Social Security Administration's computers says whether you're alive or dead. The second simplest algorithm is: combine two bits. Claude Shannon, better known as the father of information theory, was the first to realize that what transistors are doing, as they switch on and off in response to other transistors, is reasoning. (That was his master's thesis at MIT—the most important master's thesis of all time.) If transistor A turns on only when transistors B and C are both on, it's doing a tiny piece of logical reasoning. If A turns on when either B or C is on, that's another tiny logical operation. And if A turns on whenever B is off, and vice versa, that's a third operation. Believe it or not, every algorithm, no matter how complex, can be reduced to just these three operations: AND, OR, and NOT. Simple algorithms can be represented by diagrams, using different symbols for the AND, OR, and NOT operations. For example, if a fever can be caused by influenza or malaria, and you should take Tylenol for a fever and a headache, this can be expressed as follows:

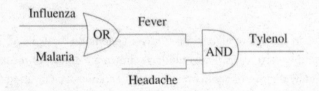

By combining many such operations, we can carry out very elaborate chains of logical reasoning. People often think computers are all about numbers, but they're not. Computers are all about logic. Numbers and arithmetic are made of logic, and so is everything else in a computer. Want to add two numbers? There's a combination of transistors that does that. Want to beat the human *Jeopardy!* champion? There's a combination of transistors for that too (much bigger, naturally).

It would be prohibitively expensive, though, if we had to build a new computer for every different thing we want to do. Rather, a modern computer is a vast assembly of transistors that can do many different things, depending on which transistors are activated. Michelangelo said that all he did was see the statue inside the block of marble and carve away the excess stone until the statue was revealed. Likewise, an algorithm carves away the excess transistors in the computer until the intended function is revealed, whether it's an airliner's autopilot or a new Pixar movie.

An algorithm is not just any set of instructions: they have to be precise and unambiguous enough to be executed by a computer. For example, a cooking recipe is not an algorithm because it doesn't exactly specify what order to do things in or exactly what each step is. Exactly how much sugar is a spoonful? As everyone who's ever tried a new recipe knows, following it may result in something delicious or a mess. In contrast, an algorithm always produces the same result. Even if a recipe specifies precisely half an ounce of sugar, we're still not out of the woods because the computer doesn't know what sugar is, or an ounce. If we wanted to program a kitchen robot to make a cake, we would have to tell it how to recognize sugar from video, how to pick up a spoon, and so on. (We're still working on that.) The computer has to know how to execute the algorithm all the way down to turning specific transistors on and off. So a cooking recipe is very far from an algorithm.

On the other hand, the following is an algorithm for playing tic-tac-toe:

If you or your opponent has two in a row, play on the remaining square.

Otherwise, if there's a move that creates two lines of two in a row, play that.

Otherwise, if the center square is free, play there.

Otherwise, if your opponent has played in a corner, play in the opposite corner.

Otherwise, if there's an empty corner, play there.

Otherwise, play on any empty square.

This algorithm has the nice property that it never loses! Of course, it's still missing many details, like how the board is represented in the computer's memory and how this representation is changed by a move. For example, we could have two bits for each square, with the value 00 if the square is empty, which changes to 01 if it has a naught and 10 if it has a cross. But it's precise and unambiguous enough that any competent programmer could fill in the blanks. It also helps that we don't really have to specify an algorithm ourselves all the way down to individual transistors; we can use preexisting algorithms as building blocks, and there's a huge number of them to choose from.

Algorithms are an exacting standard. It's often said that you don't really understand something until you can express it as an algorithm. (As Richard Feynman said, "What I cannot create, I do not understand.") Equations, the bread and butter of physicists and engineers, are really just a special kind of algorithm. For example, Newton's second law, arguably the most important equation of all time, tells you to compute the net force on an object by multiplying its mass by its acceleration. It also tells you implicitly that the acceleration is the force divided by the mass, but making that explicit is itself an algorithmic step. In any area of science, if a theory cannot be expressed as an algorithm, it's not entirely rigorous. (Not to mention you can't use a computer to solve it, which really limits what you can do with it.) Scientists make theories, and engineers make devices. Computer scientists make algorithms, which are both theories and devices.

Designing an algorithm is not easy. Pitfalls abound, and nothing can be taken for granted. Some of your intuitions will turn out to have been wrong, and you'll have to find another way. On top of designing the algorithm, you have to write it down in a language computers can understand, like Java or Python (at which point it's called a program). Then you have to debug it: find every error and fix it until the computer runs your program without screwing up. But once you have a program that does what you want, you can really go to town. Computers will do your bidding millions of times, at ultrahigh speed, without complaint. Everyone in the world can use your creation. The cost can be zero, if you so choose, or enough to make you a billionaire, if the problem you solved is important enough. A programmer—someone who creates algorithms and codes them up—is a minor god, creating universes at will. You could even say that the God of Genesis himself is a programmer: language, not manipulation, is his tool of creation. Words become worlds. Today, sitting on the couch with your laptop, you too can be a god. Imagine a universe and make it real. The laws of physics are optional.

Over time, computer scientists build on each other's work and invent algorithms for new things. Algorithms combine with other algorithms to use the results of other algorithms, in turn producing results for still more algorithms. Every second, billions of transistors in billions of computers switch billions of times. Algorithms form a new kind of ecosystem—ever growing, comparable in richness only to life itself.

Inevitably, however, there is a serpent in this Eden. It's called the complexity monster. Like the Hydra, the complexity monster has many heads. One of them is space complexity: the number of bits of information an algorithm needs to store in the computer's memory. If the algorithm needs more memory than the computer can provide, it's useless and must be discarded. Then there's the evil sister, time complexity: how long the algorithm takes to run, that is, how many steps of using and reusing the transistors it has to go through before it produces the desired results. If it's longer than we can wait, the algorithm is again useless. But the scariest face of the complexity monster is human complexity. When algorithms become too intricate for our poor human brains to understand, when the interactions between different parts of the algorithm are too many and too involved, errors creep in, we can't find them and fix them, and the algorithm doesn't do what we want. Even if we somehow make it work, it winds up being needlessly complicated for the people using it and doesn't play well with other algorithms, storing up trouble for later.

Every computer scientist does battle with the complexity monster every day. When computer scientists lose the battle, complexity seeps into our lives. You've probably noticed that many a battle has been lost. Nevertheless, we continue to build our tower of algorithms, with greater and greater difficulty. Each new generation of algorithms has to be built on top of the previous ones and has to deal with their complexities in addition to its own. The tower grows taller and taller, and it covers the whole world, but it's also increasingly fragile, like a house of cards waiting to collapse. One tiny error in an algorithm and a billion-dollar rocket explodes, or the power goes out for millions. Algorithms interact in unexpected ways, and the stock market crashes.

If programmers are minor gods, the complexity monster is the devil himself. Little by little, it's winning the war.

There has to be a better way.

Enter the learner

Every algorithm has an input and an output: the data goes into the computer, the algorithm does what it will with it, and out comes the result. Machine learning turns this around: in goes the data and the desired result and out comes the algorithm that turns one into the other. Learning algorithms—also known as learners—are algorithms that make other algorithms. With machine learning, computers write their own programs, so we don't have to.

Wow.

Computers write their own programs. Now that's a powerful idea, maybe even a little scary. If computers start to program themselves, how will we control them? Turns out we can control them quite well, as we'll see. A more immediate objection is that perhaps this sounds too good to be true. Surely writing algorithms requires intelligence, creativity, problem-solving chops—things that computers just don't have? How is machine learning distinguishable from magic? Indeed, as of today people can write many programs that computers can't learn. But, more surprisingly, computers can learn programs that people can't write. We know how to drive cars and decipher handwriting, but these skills are subconscious; we're not able to explain to a computer how to do these things. If we give a learner a sufficient number of examples of each, however, it will happily figure out how to do them on its own, at which point we can turn it loose. That's how the post office reads zip codes, and that's why self-driving cars are on the way.

The power of machine learning is perhaps best explained by a low-tech analogy: farming. In an industrial society, goods are made in factories, which means that engineers have to figure out exactly how to assemble them from their parts, how to make those parts, and so on—all the way to raw materials. It's a lot of work. Computers are the most complex goods ever invented, and designing them, the factories that make them, and the programs that run on them is a ton of work. But there's another, much older way in which we can get some of the things we need: by letting nature make them. In farming, we plant the seeds, make sure they have enough water and nutrients, and reap the grown crops. Why can't technology be more like this? It can, and that's the promise of machine learning. Learning algorithms are the seeds, data is the soil, and the learned programs are the grown plants. The machine-learning expert is like a farmer, sowing the seeds, irrigating and fertilizing the soil, and keeping an eye on the health of the crop but otherwise staying out of the way.

Once we look at machine learning this way, two things immediately jump out. The first is that the more data we have, the more we can learn. No data? Nothing to learn. Big data? Lots to learn. That's why machine learning has been turning up everywhere, driven by exponentially growing mountains of data. If machine learning was something you bought in the supermarket, its carton would say: "Just add data."

The second thing is that machine learning is a sword with which to slay the complexity monster. Given enough data, a learning program that's only a few hundred lines long can easily generate a program with millions of lines, and it can do this again and again for different problems. The reduction in complexity for the programmer is phenomenal. Of course, like the Hydra, the complexity monster sprouts new heads as soon as we cut off the old ones, but they start off smaller and take a while to grow, so we still get a big leg up.

We can think of machine learning as the inverse of programming, in the same way that the square root is the inverse of the square, or integration is the inverse of differentiation. Just as we can ask "What number squared gives 16?" or "What is the function whose derivative is $x + 1$?" we can ask, "What is the algorithm that produces this output?"

We will soon see how to turn this insight into concrete learning algorithms.

Some learners learn knowledge, and some learn skills. "All humans are mortal" is a piece of knowledge. Riding a bicycle is a skill. In machine learning, knowledge is often in the form of statistical models, because most knowledge is statistical: all humans are mortal, but only 4 percent are Americans. Skills are often in the form of procedures: if the road curves left, turn the wheel left; if a deer jumps in front of you, slam on the brakes. (Unfortunately, as of this writing Google's self-driving cars still confuse windblown plastic bags with deer.) Often, the procedures are quite simple, and it's the knowledge at their core, that's complex. If you can tell which e-mails are spam, you know which ones to delete. If you can tell how good a board position in chess is, you know which move to make (the one that leads to the best position).

Machine learning takes many different forms and goes by many different names: pattern recognition, statistical modeling, data mining, knowledge discovery, predictive analytics, data science, adaptive systems, self-organizing systems, and more. Each of these is used by different communities and has different associations. Some have a long half-life, some less so. In this book I use the term *machine learning* to refer broadly to all of them.

Machine learning is sometimes confused with artificial intelligence (or AI for short). Technically, machine learning is a subfield of AI, but it's grown so large and successful that it now eclipses its proud parent. The goal of AI is to teach computers to do what humans currently do better, and learning is arguably the most important of those things: without it, no computer can keep up with a human for long; with it, the rest follows.

In the information-processing ecosystem, learners are the superpredators. Databases, crawlers, indexers, and so on are the herbivores, patiently munging on endless fields of data. Statistical algorithms, online analytical processing, and so on are the predators. Herbivores are necessary, since without them the others couldn't exist, but superpredators have a more exciting life. A crawler is like a cow, the web is its worldwide meadow, each page is a blade of grass. When the crawler is done munging, a copy of the web is sitting on its hard disks. An indexer then makes a list of the pages where each word appears, much like the index at the end of a book. Databases, like elephants, are big and heavy and never forget. Among these patient beasts dart statistical and analytical algorithms, compacting and selecting, turning data into information. Learners eat up this information, digest it, and turn it into knowledge.

Machine-learning experts (aka machine learners) are an elite priesthood even among computer scientists. Many computer scientists, particularly those of an older generation, don't understand machine learning as well as they'd like to. This is because computer science has traditionally been all about thinking deterministically, but machine learning requires thinking statistically. If a rule for, say, labeling e-mails as spam is 99 percent accurate, that does not mean it's buggy; it may be the best you can do and good enough to be useful. This difference in thinking is a large part of why Microsoft has had a lot more trouble catching up with Google than it did with Netscape. At the end of the day, a browser is just a standard piece of software, but a search engine requires a different mind-set.

The other reason machine learners are the über-geeks is that the world has far fewer of them than it needs, even by the already dire standards of computer science. According to tech guru Tim O'Reilly, "data scientist" is the hottest job title in Silicon Valley. The McKinsey Global Institute estimates that by 2018 the United States alone will need 140,000 to 190,000 more machine-learning experts than will be available, and 1.5 million more data-savvy managers. Machine learning's applications have exploded too suddenly for education to keep up, and it has a reputation for being a difficult subject. Textbooks are liable to give you math indigestion. This difficulty is more apparent than real, however. All of the important ideas in machine learning can be expressed math-free. As you read this book, you may even find yourself inventing your own learning algorithms, with nary an equation in sight.

The Industrial Revolution automated manual work and the Information Revolution did the same for mental work, but machine learning automates automation itself. Without it, programmers become the bottleneck holding up progress. With it, the pace of progress picks up. If you're a lazy and not-too-bright computer scientist, machine learning is the ideal occupation, because learning algorithms do all the work but let you take all the credit. On the other hand, learning algorithms could put us out of our jobs, which would only be poetic justice.

By taking automation to new heights, the machine-learning revolution will cause extensive economic and social changes, just as the Internet, the personal computer, the automobile, and the steam engine did in their time. One area where these changes are already apparent is business.

Why businesses embrace machine learning

Why is Google worth so much more than Yahoo? They both make their money from showing ads on the web, and they're both top destinations. Both use auctions to sell ads and machine learning to predict how likely a user is to click on an ad (the higher the probability, the more valuable the ad). But Google's learning algorithms are much better than Yahoo's. This is not the only reason for the difference in their market caps, of course, but it's a big one. Every predicted click that doesn't happen is a wasted opportunity for the advertiser and lost revenue for the website. With Google's annual revenue of $50 billion, every 1 percent improvement in click prediction potentially means another half billion dollars in the bank, every year, for the company. No wonder Google is a big fan of machine learning, and Yahoo and others are trying hard to catch up.

Web advertising is just one manifestation of a much larger phenomenon. In every market, producers and consumers need to connect before a transaction can happen. In pre-Internet days, the main obstacles to this were physical. You could only buy books from your local bookstore, and your local bookstore had limited shelf space. But when you can download any book to your e-reader any time, the problem becomes the overwhelming number of choices. How do you browse the shelves of a bookstore that has millions of titles for sale? The same goes for other information goods: videos, music, news, tweets, blogs, plain old web pages. It also goes for every product and service that can be procured remotely: shoes, flowers, gadgets, hotel rooms, tutoring, investments. It even applies to people looking for a job or a date. How do you find each other? This is the defining problem of the Information Age, and machine learning is a big part of the solution.

As companies grow, they go through three phases. First, they do everything manually: the owners of a mom-and-pop store personally know their customers, and they order, display, and recommend items accordingly. This is nice, but it doesn't scale. In the second and least happy phase, the company grows large enough that it needs to use computers. In come the programmers, consultants, and database managers, and millions of lines of code get written to automate all the functions of the company that can be automated. Many more people are served, but not as well: decisions are made based on coarse demographic categories, and computer programs are too rigid to match humans' infinite versatility.

After a point, there just aren't enough programmers and consultants to do all that's needed, and the company inevitably turns to machine learning. Amazon can't neatly encode the tastes of all its customers in a computer program, and Facebook doesn't know how to write a program that will choose the best updates to show to each of its users. Walmart sells millions of products and has billions of choices to make every day; if the programmers at Walmart tried to write a program to make all of them, they would never be done. Instead, what these companies do is turn learning algorithms loose on the mountains of data they've accumulated and let them divine what customers want.

Learning algorithms are the matchmakers: they find producers and consumers for each other, cutting through the information overload. If they're smart enough, you get the best of both worlds: the vast choice and low cost of the large scale, with the personalized touch of the small. Learners are not perfect, and the last step of the decision is usually still for humans to make, but learners intelligently reduce the choices to something a human can manage.

In retrospect, we can see that the progression from computers to the Internet to machine learning was inevitable: computers enable the Internet, which creates a flood of data and the problem of limitless choice; and machine learning uses the flood of data to help solve the limitless choice problem. The Internet by itself is not enough to move demand from "one size fits all" to the long tail of infinite variety. Netflix may have one hundred thousand DVD titles in stock, but if customers don't know how to find the ones they like, they will default to choosing the hits. It's only when Netflix has a learning algorithm to figure out your tastes and recommend DVDs that the long tail really takes off.

Once the inevitable happens and learning algorithms become the middlemen, power becomes concentrated in them. Google's algorithms largely determine what information you find, Amazon's what products you buy, and Match.com's who you date. The last mile is still yours—choosing from among the options the algorithms present you with—but 99.9 percent of the selection was done by them. The success or failure of a company now depends on how much the learners like its products, and the success of a whole economy—whether everyone gets the best products for their needs at the best price—depends on how good the learners are.

The best way for a company to ensure that learners like its products is to run them itself. Whoever has the best algorithms and the most data wins. A new type of network effect takes hold: whoever has the most customers accumulates the most data, learns the best models, wins the most new customers, and so on in a virtuous circle (or a vicious one, if you're the competition). Switching from Google to Bing may be easier than switching from Windows to Mac, but in practice you don't because Google, with its head start and larger market share, knows better what you want, even if Bing's technology is just as good. And pity a new entrant into the search business, starting with zero data against engines with over a decade of learning behind them.

You might think that after a while more data is just more of the same, but that saturation point is nowhere in sight. The long tail keeps going. If you look at the recommendations Amazon or Netflix gives you, it's clear they're still very crude, and Google's search results still leave a lot to be desired. Every feature of a product, every corner of a website can potentially be improved using machine learning. Should the link at the bottom of a page be red or blue? Try them both and see which one gets the most clicks. Better still, keep the learners running and continuously adjust all aspects of the website.

The same dynamic happens in any market where there's lots of choice and lots of data. The race is on, and whoever learns fastest wins. It doesn't stop with understanding customers better: companies can apply machine learning to every aspect of their operations, provided data is available, and data is pouring in from computers, communication devices, and ever-cheaper and more ubiquitous sensors. "Data is the new oil" is a popular refrain, and as with oil, refining it is big business. IBM, as well plugged into the corporate world as anyone, has organized its growth strategy around providing analytics to companies. Businesses look at data as a strategic asset: What data do I have that my competitors don't? How can I take advantage of it? What data do my competitors have that I don't?

In the same way that a bank without databases can't compete with a bank that has them, a company without machine learning can't keep up with one that uses it. While the first company's experts write a thousand rules to predict what its customers want, the second company's algorithms learn billions of rules, a whole set of them for each individual customer. It's about as fair as spears against machine guns. Machine learning is a cool new technology, but that's not why businesses embrace it. They embrace it because they have no choice.

Supercharging the scientific method

Machine learning is the scientific method on steroids. It follows the same process of generating, testing, and discarding or refining hypotheses. But while a scientist may spend his or her whole life coming up with and testing a few hundred hypotheses, a machine-learning system can do the same in a fraction of a second. Machine learning automates discovery. It's no surprise, then, that it's revolutionizing science as much as it's revolutionizing business.

To make progress, every field of science needs to have data commensurate with the complexity of the phenomena it studies. This is why physics was the first science to take off: Tycho Brahe's recordings of the positions of the planets and Galileo's observations of pendulums and inclined planes were enough to infer Newton's laws. It's also why molecular biology, despite being younger than neuroscience, has outpaced it: DNA microarrays and high-throughput sequencing provide a volume of data that neuroscientists can only hope for. And it's the reason why social science research is such an uphill battle: if all you have is a sample of a hundred people, with a dozen measurements apiece, all you can model is some very narrow phenomenon. But even this narrow phenomenon does not exist in isolation; it's affected by a myriad others, which means you're still far from understanding it.

The good news today is that sciences that were once data-poor are now data-rich. Instead of paying fifty bleary-eyed undergraduates to perform some task in the lab, psychologists can get as many subjects as they want by posting the task on Amazon's Mechanical Turk. (It makes for a more diverse sample too.) It's getting hard to remember, but little more than a decade ago sociologists studying social networks lamented that they couldn't get their hands on a network with more than a few hundred members. Now there's Facebook, with over a billion. A good chunk of those members post almost blow-by-blow accounts of their lives too; it's like having a live feed of social life on planet Earth. In neuroscience, connectomics and functional magnetic resonance imaging have opened an extraordinarily detailed window into the brain. In molecular biology, databases of genes and proteins grow exponentially. Even in "older" sciences like physics and astronomy, progress continues because of the flood of data pouring forth from particle accelerators and digital sky surveys.

Big data is no use if you can't turn it into knowledge, however, and there aren't enough scientists in the world for the task. Edwin Hubble discovered new galaxies by poring over photographic plates, but you can bet the half-billion sky objects in the Sloan Digital Sky Survey weren't identified that way. It would be like trying to count the grains of sand on a beach by hand. You can write rules to distinguish galaxies from stars from noise objects (such as birds, planes, Superman), but they're not very accurate. Instead, the SKICAT (sky image cataloging and analysis tool) project used a learning algorithm. Starting from plates where objects were labeled with the correct categories, it figured out what characterizes each one and applied the result to all the unlabeled plates. Even better, it could classify objects that were too faint for humans to label, and these comprise the majority of the survey.

With big data and machine learning, you can understand much more complex phenomena than before. In most fields, scientists have traditionally used only very limited kinds of models, like linear regression, where the curve you fit to the data is always a straight line. Unfortunately, most phenomena in the world are nonlinear. (Or fortunately, since otherwise life would be very boring—in fact, there would be no life.) Machine learning opens up a vast new world of nonlinear models. It's like turning on the lights in a room where only a sliver of moonlight filtered before.

In biology, learning algorithms figure out where genes are located in a DNA molecule, where superfluous bits of RNA get spliced out before proteins are synthesized, how proteins fold into their characteristic shapes, and how different conditions affect the expression of different genes. Rather than testing thousands of new drugs in the lab, learners predict whether they will work, and only the most promising get tested. They also weed out molecules likely to have nasty side effects, like cancer. This avoids expensive failures, like candidate drugs being nixed only after human trials have begun.

The biggest challenge, however, is assembling all this information into a coherent whole. What are all the things that affect your risk of heart disease, and how do they interact? All Newton needed was three laws of motion and one of gravitation, but a complete model of a cell, an organism, or a society is more than any one human can discover. As knowledge grows, scientists specialize ever more narrowly, but no one is able to put the pieces together because there are far too many pieces. Scientists collaborate, but language is a very slow medium of communication. Scientists try to keep up with others' research, but the volume of publications is so high that they fall farther and farther behind. Often, redoing an experiment is easier than finding the paper that reported it. Machine learning comes to the rescue, scouring the literature for relevant information, translating one area's jargon into another's, and even making connections that scientists weren't aware of. Increasingly, machine learning acts as a giant hub, through which modeling techniques invented in one field make their way into others.

If computers hadn't been invented, science would have ground to a halt in the second half of the twentieth century. This might not have been immediately apparent to the scientists because they would have been focused on whatever limited progress they could still make, but the ceiling for that progress would have been much, much lower. Similarly, without machine learning, many sciences would face diminishing returns in the decades to come.

To see the future of science, take a peek inside a lab at the Manchester Institute of Biotechnology, where a robot by the name of Adam is hard at work figuring out which genes encode which enzymes in yeast. Adam has a model of yeast metabolism and general knowledge of genes and proteins. It makes hypotheses, designs experiments to test them, physically carries them out, analyzes the results, and comes up with new hypotheses until it's satisfied. Today, human scientists still independently check Adam's conclusions before they believe them, but tomorrow they'll leave it to robot scientists to check each other's hypotheses.

A billion Bill Clintons

Machine learning was the kingmaker in the 2012 presidential election. The factors that usually decide presidential elections—the economy, likability of the candidates, and so on—added up to a wash, and the outcome came down to a few key swing states. Mitt Romney's campaign followed a conventional polling approach, grouping voters into broad categories and targeting each one or not. Neil Newhouse, Romney's pollster, said that "if we can win independents in Ohio, we can win this race." Romney won them by 7 percent but still lost the state and the election.

In contrast, President Obama hired Rayid Ghani, a machine-learning expert, as chief scientist of his campaign, and Ghani proceeded to put together the greatest analytics operation in the history of politics. They consolidated all voter information into a single database; combined it with what they could get from social networking, marketing, and other sources; and set about predicting four things for each individual voter: how likely he or she was to support Obama, show up at the polls, respond to the campaign's reminders to do so, and change his or her mind about the election based on a conversation about a specific issue. Based on these voter models, every night the campaign ran 66,000 simulations of the election and used the results to direct its army of volunteers: whom to call, which doors to knock on, what to say.

In politics, as in business and war, there is nothing worse than seeing your opponent make moves that you don't understand and don't know what to do about until it's too late. That's what happened to the Romney campaign. They could see the other side buying ads in particular cable stations in particular towns but couldn't tell why; their crystal ball was too fuzzy. In the end, Obama won every battleground state save North Carolina and by larger margins than even the most accurate pollsters had predicted. The most accurate pollsters, in turn, were the ones (like Nate Silver) who used the most sophisticated prediction techniques; they were less accurate than the Obama campaign because they had fewer resources. But they were a lot more accurate than the traditional pundits, whose predictions were based on their expertise.

You might think the 2012 election was a fluke: most elections are not close enough for machine learning to be the deciding factor. But machine learning will *cause* more elections to be close in the future. In politics, as in everything, learning is an arms race. In the days of Karl Rove, a former direct marketer and data miner, the Republicans were ahead.

By 2012, they'd fallen behind, but now they're catching up again. We don't know who'll be ahead in the next election cycle, but both parties will be working hard to win. That means understanding the voters better and tailoring the candidates' pitches—even choosing the candidates themselves—accordingly. The same applies to entire party platforms, during and between election cycles: if detailed voter models, based on hard data, say a party's current platform is a losing one, the party will change it. As a result, major events aside, gaps between candidates in the polls will be smaller and shorter lived. Other things being equal, the candidates with the better voter models will win, and voters will be better served for it.

One of the greatest talents a politician can have is the ability to understand voters, individually or in small groups, and speak directly to them (or seem to). Bill Clinton is the paradigmatic example of this in recent memory. The effect of machine learning is like having a dedicated Bill Clinton for every voter. Each of these mini-Clintons is a far cry from the real one, but they have the advantage of numbers; even Bill Clinton can't know what every single voter in America is thinking (although he'd surely like to). Learning algorithms are the ultimate retail politicians.

Of course, as with companies, politicians can put their machine-learned knowledge to bad uses as well as good ones. For example, they could make inconsistent promises to different voters. But voters, media, and watchdog organizations can do their own data mining and expose politicians who cross the line. The arms race is not just between candidates but among all participants in the democratic process.

The larger outcome is that democracy works better because the bandwidth of communication between voters and politicians increases enormously. In these days of high-speed Internet, the amount of information your elected representatives get from you is still decidedly nineteenth century: a hundred bits or so every two years, as much as fits on a ballot. This is supplemented by polling and perhaps the occasional e-mail or town-hall meeting, but that's still precious little. Big data and machine learning change the equation. In the future, provided voter models are accurate, elected officials will be able to ask voters what they want a thousand times a day and act accordingly—without having to pester the actual flesh-and-blood citizens.

One if by land, two if by Internet

Out in cyberspace, learning algorithms man the nation's ramparts. Every day, foreign attackers attempt to break into computers at the Pentagon, defense contractors, and other companies and government agencies. Their tactics change continually; what worked against yesterday's attacks is powerless against today's. Writing code to detect and block each one would be as effective as the Maginot Line, and the Pentagon's Cyber Command knows it. But machine learning runs into a problem if an attack is the first of its kind and there aren't any previous examples of it to learn from. Instead, learners build models of normal behavior, of which there's plenty, and flag anomalies. Then they call in the cavalry (aka system administrators). If cyberwar ever comes to pass, the generals will be human, but the foot soldiers will be algorithms. Humans are too slow and too few and would be quickly swamped by an army of bots. We need our own bot army, and machine learning is like West Point for bots.

Cyberwar is an instance of asymmetric warfare, where one side can't match the other's conventional military power but can still inflict grievous damage. A handful of terrorists armed with little more than box cutters can knock down the Twin Towers and kill thousands of innocents. All the biggest threats to US security today are in the realm of asymmetric warfare, and there's an effective weapon against all of them: information. If the enemy can't hide, he can't survive. The good news is that we have plenty of information, and that's also the bad news.

The National Security Agency (NSA) has become infamous for its bottomless appetite for data: by one estimate, every day it intercepts over a billion phone calls and other communications around the globe. Privacy issues aside, however, it doesn't have millions of staffers to eavesdrop on all these calls and e-mails or even just keep track of who's talking to whom. The vast majority of calls are perfectly innocent, and writing a program to pick out the few suspicious ones is very hard. In the old days, the NSA used keyword matching, but that's easy to get around. (Just call the bombing a "wedding" and the bomb the "wedding cake.") In the twenty-first century, it's a job for machine learning. Secrecy is the NSA's trademark, but its director has testified to Congress that mining of phone logs has already halted dozens of terrorism threats.

Terrorists can hide in the crowd at a football game, but learners can pick out their faces. They can make exotic bombs, but learners can sniff them out. Learners can also do something more subtle: connect the dots between events that individually seem harmless but add up to an ominous pattern. This approach could have prevented 9/11. There's a further twist: once a learned program is deployed, the bad guys change their behavior to defeat it. This contrasts with the natural world, which always works the same way. The solution is to marry machine learning with game theory, something I've worked on in the past: don't just learn to defeat what your opponent does now; learn to parry what he might do against your learner. Factoring in the costs and benefits of different actions, as game theory does, can also help strike the right balance between privacy and security.

During the Battle of Britain, the Royal Air Force held back the Luftwaffe despite being heavily outnumbered. German pilots couldn't understand how, wherever they went, they always ran into the RAF. The British had a secret weapon: radar, which detected the German planes well before they crossed into Britain's airspace. Machine learning is like having a radar that sees into the future. Don't just react to your adversary's moves; predict them and preempt them.

An example of this closer to home is what's known as predictive policing. By forecasting crime trends and strategically focusing patrols where they're most likely to be needed, as well as taking other preventive measures, a city's police force can effectively do the job of a much larger one. In many ways, law enforcement is similar to asymmetric warfare, and many of the same learning techniques apply, whether it's in fraud detection, uncovering criminal networks, or plain old beat policing.

Machine learning also has a growing role on the battlefield. Learners can help dissipate the fog of war, sifting through reconnaissance imagery, processing after-action reports, and piecing together a picture of the situation for the commander. Learning powers the brains of military robots, helping them keep their bearings, adapt to the terrain, distinguish enemy vehicles from civilian ones, and home in on their targets. DARPA's AlphaDog carries soldiers' gear for them. Drones can fly autonomously with the help of learning algorithms; although they are still partly controlled by human pilots, the trend is for one pilot to oversee larger and larger swarms. In the army of the future, learners will greatly outnumber soldiers, saving countless lives.

Where are we headed?

Technology trends come and go all the time. What's unusual about machine learning is that, through all these changes, through boom and bust, it just keeps growing. Its first big hit was in finance, predicting stock ups and downs, starting in the late 1980s. The next wave was mining corporate databases, which by the mid-1990s were starting to grow quite large, and in areas like direct marketing, customer relationship management, credit scoring, and fraud detection. Then came the web and e-commerce, where automated personalization quickly became de rigueur. When the dot-com bust temporarily curtailed that, the use of learning for web search and ad placement took off. For better or worse, the 9/11 attacks put machine learning in the front line of the war on terror. Web 2.0 brought a swath of new applications, from mining social networks to figuring out what bloggers are saying about your products. In parallel, scientists of all stripes were increasingly turning to large-scale modeling, with molecular biologists and astronomers leading the charge. The housing bust barely registered; its main effect was a welcome transfer of talent from Wall Street to Silicon Valley. In 2011, the "big data" meme hit, putting machine learning squarely in the center of the global economy's future. Today, there seems to be hardly an area of human endeavor untouched by machine learning, including seemingly unlikely candidates like music, sports, and wine tasting.

As remarkable as this growth is, it's only a foretaste of what's to come. Despite its usefulness, the generation of learning algorithms currently at work in industry is, in fact, quite limited. When the algorithms now in the lab make it to the front lines, Bill Gates's remark that a breakthrough in machine learning would be worth ten Microsofts will seem conservative. And if the ideas that *really* put a glimmer in researchers' eyes bear fruit, machine learning will bring about not just a new era of civilization, but a new stage in the evolution of life on Earth.

What makes this possible? How do learning algorithms work? What can't they currently do, and what will the next generation look like? How will the machine-learning revolution unfold? And what opportunities and dangers should you look out for? That's what this book is about—read on!

CHAPTER TWO

The Master Algorithm

Even more astonishing than the breadth of applications of machine learning is that it's the *same* algorithms doing all of these different things. Outside of machine learning, if you have two different problems to solve, you need to write two different programs. They might use some of the same infrastructure, like the same programming language or the same database system, but a program to, say, play chess is of no use if you want to process credit-card applications. In machine learning, the same algorithm can do both, provided you give it the appropriate data to learn from. In fact, just a few algorithms are responsible for the great majority of machine-learning applications, and we'll take a look at them in the next few chapters.

For example, consider Naïve Bayes, a learning algorithm that can be expressed as a single short equation. Given a database of patient records—their symptoms, test results, and whether or not they had some particular condition—Naïve Bayes can learn to diagnose the condition in a fraction of a second, often better than doctors who spent many years in medical school. It can also beat medical expert systems that took thousands of person-hours to build. The same algorithm is widely used to learn spam filters, a problem that at first sight has nothing to do with medical diagnosis. Another simple learner, called the nearest-neighbor algorithm, has been used for everything from handwriting recognition to controlling robot hands to recommending books and movies you might like. And decision tree learners are equally apt at deciding whether your credit-card application should be accepted, finding splice junctions in DNA, and choosing the next move in a game of chess.

Not only can the same learning algorithms do an endless variety of different things, but they're shockingly simple compared to the algorithms they replace. Most learners can be coded up in a few hundred lines, or perhaps a few thousand if you add a lot of bells and whistles. In contrast, the programs they replace can run in the hundreds of thousands or even millions of lines, and a single learner can induce an unlimited number of different programs.

If so few learners can do so much, the logical question is: Could one learner do everything? In other words, could a single algorithm learn all that can be learned from data? This is a very tall order, since it would ultimately include everything in an adult's brain, everything evolution has created, and the sum total of all scientific knowledge. But in fact all the major learners—including nearest-neighbor, decision trees, and Bayesian networks, a generalization of Naïve Bayes—are universal in the following sense: if you give the learner enough of the appropriate data, it can approximate any function arbitrarily closely—which is math-speak for learning anything. The catch is that "enough data" could be infinite. Learning from finite data requires making assumptions, as we'll see, and different learners make different assumptions, which makes them good for some things but not others.

But what if instead of leaving these assumptions embedded in the algorithm we make them an explicit input, along with the data, and allow the user to choose which ones to plug in, perhaps even state new ones? Is there an algorithm that can take in any data and assumptions and output the knowledge that's implicit in them? I believe so. Of course, we have to put some limits on what the assumptions can be, otherwise we could cheat by giving the algorithm the entire target knowledge, or close to it, in the form of assumptions. But there are many ways to do this, from limiting the size of the input to requiring that the assumptions be no stronger than those of current learners.

The question then becomes: How weak can the assumptions be and still allow all relevant knowledge to be derived from finite data? Notice the word *relevant*: we're only interested in knowledge about our world, not about worlds that don't exist. So inventing a universal learner boils down to discovering the deepest regularities in our universe, those that all phenomena share, and then figuring out a computationally efficient way to combine them with data. This requirement of computational efficiency precludes just using the laws of physics as the regularities, as we'll see. It does not, however, imply that the universal learner has to be as efficient as more specialized ones. As so often happens in computer science, we're willing to sacrifice efficiency for generality. This also applies to the amount of data required to learn a given target knowledge: a universal learner will generally need more data than a specialized one, but that's OK provided we have the necessary amount—and the bigger data gets, the more likely this will be the case.

Here, then, is the central hypothesis of this book:

All knowledge—past, present, and future—can be derived from data by a single, universal learning algorithm.

I call this learner the Master Algorithm. If such an algorithm is possible, inventing it would be one of the greatest scientific achievements of all time. In fact, the Master Algorithm is the last thing we'll ever have to invent because, once we let it loose, it will go on to invent everything else that can be invented. All we need to do is provide it with enough of the right kind of data, and it will discover the corresponding knowledge. Give it a video stream, and it learns to see. Give it a library, and it learns to read. Give it the results of physics experiments, and it discovers the laws of physics. Give it DNA crystallography data, and it discovers the structure of DNA.

This may sound far-fetched: How could one algorithm possibly learn so many different things and such difficult ones? But in fact many lines of evidence point to the existence of a Master Algorithm. Let's see what they are.

The argument from neuroscience

In April 2000, a team of neuroscientists from MIT reported in *Nature* the results of an extraordinary experiment. They rewired the brain of a ferret, rerouting the connections from the eyes to the auditory cortex (the part of the brain responsible for processing sounds) and rerouting the connections from the ears to the visual cortex. You'd think the result would be a severely disabled ferret, but no: the auditory cortex learned to see, the visual cortex learned to hear, and the ferret was fine. In normal mammals, the visual cortex contains a map of the retina: neurons connected to nearby regions of the retina are close to each other in the cortex. Instead, the rewired ferrets developed a map of the retina in the auditory cortex. If the visual input is redirected instead to the somatosensory cortex, responsible for touch perception, it too learns to see. Other mammals also have this ability.

In congenitally blind people, the visual cortex can take over other brain functions. In deaf ones, the auditory cortex does the same. Blind people can learn to "see" with their tongues by sending video images from a head-mounted camera to an array of electrodes placed on the tongue, with high voltages corresponding to bright pixels and low voltages to dark ones. Ben Underwood was a blind kid who taught himself to use echolocation to navigate, like bats do. By clicking his tongue and listening to the echoes, he could walk around without bumping into obstacles, ride a skateboard, and even play basketball. All of this is evidence that the brain uses the same learning algorithm throughout, with the areas dedicated to the different senses distinguished only by the different inputs they are connected to (e.g., eyes, ears, nose). In turn, the associative areas acquire their function by being connected to multiple sensory regions, and the "executive" areas acquire theirs by connecting the associative areas and motor output.

Examining the cortex under a microscope leads to the same conclusion. The same wiring pattern is repeated everywhere. The cortex is organized into columns with six distinct layers, feedback loops running to another brain structure called the thalamus, and a recurring pattern of short-range inhibitory connections and longer-range excitatory ones. A certain amount of variation is present, but it looks more like different parameters or settings of the same algorithm than different algorithms. Low-level sensory areas have more noticeable differences, but as the rewiring experiments show, these are not crucial. The cerebellum, the evolutionarily older part of the brain responsible for low-level motor control, has a clearly different and very regular architecture, built out of much smaller neurons, so it would seem that at least motor learning uses a different algorithm. If someone's cerebellum is injured, however, the cortex takes over its function. Thus it seems that evolution kept the cerebellum around not because it does something the cortex can't, but just because it's more efficient.

The computations taking place within the brain's architecture are also similar throughout. All information in the brain is represented in the same way, via the electrical firing patterns of neurons. The learning mechanism is also the same: memories are formed by strengthening the connections between neurons that fire together, using a biochemical process known as long-term potentiation. All this is not just true of humans: different animals have similar brains. Ours is unusually large, but seems to be built along the same principles as other animals'.

Another line of argument for the unity of the cortex comes from what might be called the poverty of the genome. The number of connections in your brain is over a million times the number of letters in your genome, so it's not physically possible for the genome to specify in detail how the brain is wired.

The most important argument for the brain being the Master Algorithm, however, is that it's responsible for everything we can perceive and imagine. If something exists but the brain can't learn it, we don't know it exists. We may just not see it or think it's random. Either way, if we implement the brain in a computer, that algorithm can learn everything we can. Thus one route—arguably the most popular one—to inventing the Master Algorithm is to reverse engineer the brain. Jeff Hawkins took a stab at this in his book *On Intelligence*. Ray Kurzweil pins his hopes for the Singularity—the rise of artificial intelligence that greatly exceeds the human variety—on doing just that and takes a stab at it himself in his book *How to Create a Mind*. Nevertheless, this is only one of several possible approaches, as we'll see. It's not even necessarily the most promising one, because the brain is phenomenally complex, and we're still in the very early stages of deciphering it. On the other hand, if we can't figure out the Master Algorithm, the Singularity won't happen any time soon.

Not all neuroscientists believe in the unity of the cortex; we need to learn more before we can be sure. The question of just what the brain can and can't learn is also hotly debated. But if there's something we know but the brain can't learn, it must have been learned by evolution.

The argument from evolution

Life's infinite variety is the result of a single mechanism: natural selection. Even more remarkable, this mechanism is of a type very familiar to computer scientists: iterative search, where we solve a problem by trying many candidate solutions, selecting and modifying the best ones, and repeating these steps as many times as necessary. Evolution *is* an algorithm. Paraphrasing Charles Babbage, the Victorian-era computer pioneer, God created not species but the algorithm for creating species. The "endless forms most beautiful" Darwin spoke of in the conclusion of *The Origin of Species* belie a most beautiful unity: all of those forms are encoded in strings of DNA, and all of them come about by modifying and combining those strings. Who would have guessed, given only a description of this algorithm, that it could produce you and me? If evolution can learn us, it can conceivably also learn everything that can be learned, provided we implement it on a powerful enough computer. Indeed, evolving programs by simulating natural selection is a popular endeavor in machine learning. Evolution, then, is another promising path to the Master Algorithm.

Evolution is the ultimate example of how much a simple learning algorithm can achieve given enough data. Its input is the experience and fate of all living creatures that ever existed. (Now *that's* big data.) On the other hand, it's been running for over three billion years on the most powerful computer on Earth: Earth itself. A computer version of it had better be faster and less data intensive than the original. Which one is the better model for the Master Algorithm: evolution or the brain? This is machine learning's version of the nature versus nurture debate. And, just as nature and nurture combine to produce us, perhaps the true Master Algorithm contains elements of both.

The argument from physics

In a famous 1959 essay, the physicist and Nobel laureate Eugene Wigner marveled at what he called "the unreasonable effectiveness of mathematics in the natural sciences." By what miracle do laws induced from scant observations turn out to apply far beyond them? How can the laws be many orders of magnitude more precise than the data they are based on? Most of all, why is it that the simple, abstract language of mathematics can accurately capture so much of our infinitely complex world? Wigner considered this a deep mystery, in equal parts fortunate and unfathomable. Nevertheless, it is so, and the Master Algorithm is a logical extension of it.

If the world were just a blooming, buzzing confusion, there would be reason to doubt the existence of a universal learner. But if everything we experience is the product of a few simple laws, then it makes sense that a single algorithm can induce all that can be induced. All the Master Algorithm has to do is provide a shortcut to the laws' consequences, replacing impossibly long mathematical derivations with much shorter ones based on actual observations.

For example, we believe that the laws of physics gave rise to evolution, but we don't know how. Instead, we can induce natural selection directly from observations, as Darwin did. Countless wrong inferences could be drawn from those observations, but most of them never occur to us, because our inferences are influenced by our broad knowledge of the world, and that knowledge is consistent with the laws of nature.

How much of the character of physical law percolates up to higher domains like biology and sociology remains to be seen, but the study of chaos provides many tantalizing examples of very different systems with similar behavior, and the theory of universality explains them. The Mandelbrot set is a beautiful example of how a very simple iterative procedure can give rise to an inexhaustible variety of forms. If the mountains, rivers, clouds, and trees of the world are all the result of such procedures—and fractal geometry shows they are—perhaps those procedures are just different parametrizations of a single one that we can induce from them.

In physics, the same equations applied to different quantities often describe phenomena in completely different fields, like quantum mechanics, electromagnetism, and fluid dynamics. The wave equation, the diffusion equation, Poisson's equation: once we discover it in one field, we can more readily discover it in others; and once we've learned how to solve it in one field, we know how to solve it in all. Moreover, all these equations are quite simple and involve the same few derivatives of quantities with respect to space and time. Quite conceivably, they are all instances of a master equation, and all the Master Algorithm needs to do is figure out how to instantiate it for different data sets.

Another line of evidence comes from optimization, the branch of mathematics concerned with finding the input to a function that produces its highest output. For example, finding the sequence of stock purchases and sales that maximizes your total returns is an optimization problem. In optimization, simple functions often give rise to surprisingly complex solutions. Optimization plays a prominent role in almost every field of science, technology, and business, including machine learning. Each field optimizes within the constraints defined by optimizations in other fields. We try to maximize our happiness within economic constraints, which are firms' best solutions within the constraints of the available technology—which in turn consists of the best solutions we could find within the constraints of biology and physics. Biology, in turn, is the result of optimization by evolution within the constraints of physics and chemistry, and the laws of physics themselves are solutions to optimization problems. Perhaps, then, everything that exists is the progressive solution of an overarching optimization problem, and the Master Algorithm follows from the statement of that problem.

Physicists and mathematicians are not the only ones who find unexpected connections between disparate fields. In his book *Consilience*, the distinguished biologist E. O. Wilson makes an impassioned argument for the unity of all knowledge, from science to the humanities. The Master Algorithm is the ultimate expression of this unity: if all knowledge shares a common pattern, the Master Algorithm exists, and vice versa.

Nevertheless, physics is unique in its simplicity. Outside physics and engineering, the track record of mathematics is more mixed. Sometimes it's only reasonably effective, and sometimes its models are too oversimplified to be useful. This tendency to oversimplify stems from the limitations of the human mind, however, not from the limitations of mathematics. Most of the brain's hardware (or rather, wetware) is devoted to sensing and moving, and to do math we have to borrow parts of it that evolved for language. Computers have no such limitations and can easily turn big data into very complex models. Machine learning is what you get when the unreasonable effectiveness of mathematics meets the unreasonable effectiveness of data. Biology and sociology will never be as simple as physics, but the method by which we discover their truths can be.

The argument from statistics

According to one school of statisticians, a single simple formula underlies all learning. Bayes' theorem, as the formula is known, tells you how to update your beliefs whenever you see new evidence. A Bayesian learner starts with a set of hypotheses about the world. When it sees a new piece of data, the hypotheses that are compatible with it become more likely, and the hypotheses that aren't become less likely (or even impossible). After seeing enough data, a single hypothesis dominates, or a few do. For example, if I'm looking for a program that accurately predicts stock movements and a stock that a candidate program had predicted would go up instead goes down, that candidate loses credibility. After I've reviewed a number of candidates, only a few credible ones will remain, and they will encapsulate my new knowledge of the stock market.

Bayes' theorem is a machine that turns data into knowledge. According to Bayesian statisticians, it's the *only* correct way to turn data into knowledge. If they're right, either Bayes' theorem is the Master Algorithm or it's the engine that drives it. Other statisticians have serious reservations about the way Bayes' theorem is used and prefer different ways to learn from data. In the days before computers, Bayes' theorem could only be applied to very simple problems, and the idea of it as a universal learner would have seemed far-fetched. With big data and big computing to go with it, however, Bayes can find its way in vast hypothesis spaces and has spread to every conceivable field of knowledge. If there's a limit to what Bayes can learn, we haven't found it yet.

The argument from computer science

When I was a senior in college, I wasted a summer playing Tetris, a highly addictive video game where variously shaped pieces fall from above and which you try to pack as closely together as you can; the game is over when the pile of pieces reaches the top of the screen. Little did I know that this was my introduction to NP-completeness, the most important problem in theoretical computer science. Turns out that, far from an idle pursuit, mastering Tetris—*really* mastering it—is one of the most useful things you could ever do. If you can solve Tetris, you can solve thousands of the hardest and most important problems in science, technology, and management—all in one fell swoop. That's because at heart they are all the *same* problem. This is one of the most astonishing facts in all of science.

Figuring out how proteins fold into their characteristic shapes; reconstructing the evolutionary history of a set of species from their DNA; proving theorems in propositional logic; detecting arbitrage opportunities in markets with transaction costs; inferring a three-dimensional shape from two-dimensional views; compressing data on a disk; forming a stable coalition in politics; modeling turbulence in sheared flows; finding the safest portfolio of investments with a given return, the shortest route to visit a set of cities, the best layout of components on a microchip, the best placement of sensors in an ecosystem, or the lowest energy state of a spin glass; scheduling flights, classes, and factory jobs; optimizing resource allocation, urban traffic flow, social welfare, and (most important) your Tetris score: these are all NP-complete problems, meaning that if you can efficiently solve one of them you can efficiently solve all problems in the class NP, including each other. Who would have guessed that all these problems, superficially so different, are really the same? But if they are, it makes sense that one algorithm could learn to solve all of them (or, more precisely, all efficiently solvable instances).

P and NP are the two most important classes of problems in computer science. (The names are not very mnemonic, unfortunately.) A problem is in P if we can solve it efficiently, and it's in NP if we can efficiently check its solution. The famous P = NP question is whether every efficiently checkable problem is also efficiently solvable. Because of NP-completeness, all it takes to answer it is to prove that *one* NP-complete problem is efficiently solvable (or not). NP is not the hardest class of problems in computer science, but it's arguably the hardest "realistic" class: if you can't even check a problem's solution before the universe ends, what's the point of trying to solve it? Humans are good at solving NP problems approximately, and conversely, problems that we find interesting (like Tetris) often have an "NP-ness" about them. One definition of artificial intelligence is that it consists of finding heuristic solutions to NP-complete problems. Often, we do this by reducing them to satisfiability, the canonical NP-complete problem: Can a given logical formula ever be true, or is it self-contradictory? If we invent a learner that can learn to solve satisfiability, it has a good claim to being the Master Algorithm.

NP-completeness aside, the sheer existence of computers is itself a powerful sign that there is a Master Algorithm. If you could travel back in time to the early twentieth century and tell people that a soon-to-be-invented machine would solve problems in every realm of human endeavor—the *same* machine for *every* problem—no one would believe you. They would say that each machine can only do one thing: sewing machines don't type, and typewriters don't sew. Then in 1936 Alan Turing imagined a curious contraption with a tape and a head that read and wrote symbols on it, now known as a Turing machine. Every conceivable problem that can be solved by logical deduction can be solved by a Turing machine. Furthermore, a so-called universal Turing machine can simulate any other by reading its specification from the tape—in other words, it can be programmed to do anything.

The Master Algorithm is for induction, the process of learning, what the Turing machine is for deduction. It can learn to simulate any other algorithm by reading examples of its input-output behavior. Just as there are many models of computation equivalent to a Turing machine, there are probably many different equivalent formulations of a universal learner. The point, however, is to find the first such formulation, just as Turing found the first formulation of the general-purpose computer.

Machine learners versus knowledge engineers

Of course, the Master Algorithm has at least as many skeptics as it has proponents. Doubt is in order when something looks like a silver bullet. The most determined resistance comes from machine learning's perennial foe: knowledge engineering. According to its proponents, knowledge can't be learned automatically; it must be programmed into the computer by human experts. Sure, learners can extract some things from data, but nothing you'd confuse with *real* knowledge. To knowledge engineers, big data is not the new oil; it's the new snake oil.

In the early days of AI, machine learning seemed like the obvious path to computers with humanlike intelligence; Turing and others thought it was the *only* plausible path. But then the knowledge engineers struck back, and by 1970 machine learning was firmly on the back burner. For a moment in the 1980s, it seemed like knowledge engineering was about to take over the world, with companies and countries making massive investments in it. But disappointment soon set in, and machine learning began its inexorable rise, at first quietly, and then riding a roaring wave of data.

Despite machine learning's successes, the knowledge engineers remain unconvinced. They believe that its limitations will soon become apparent, and the pendulum will swing back. Marvin Minsky, an MIT professor and AI pioneer, is a prominent member of this camp. Minsky is not just skeptical of machine learning as an alternative to knowledge engineering, he's skeptical of *any* unifying ideas in AI. Minsky's theory of intelligence, as expressed in his book *The Society of Mind*, could be unkindly characterized as "the mind is just one damn thing after another." *The Society of Mind* is a laundry list of hundreds of separate ideas, each with its own vignette. The problem with this approach to AI is that it doesn't work; it's stamp collecting by computer. Without machine learning, the number of ideas needed to build an intelligent agent is infinite. If a robot had all the same capabilities as a human except learning, the human would soon leave it in the dust.

Minsky was an ardent supporter of the Cyc project, the most notorious failure in the history of AI. The goal of Cyc was to solve AI by entering into a computer all the necessary knowledge. When the project began in the 1980s, its leader, Doug Lenat, confidently predicted success within a decade. Thirty years later, Cyc continues to grow without end in sight, and commonsense reasoning still eludes it. Ironically, Lenat has belatedly embraced populating Cyc by mining the web, not because Cyc can read, but because there's no other way.

Even if by some miracle we managed to finish coding up all the necessary pieces, our troubles would be just beginning. Over the years, a number of research groups have attempted to build complete intelligent agents by putting together algorithms for vision, speech recognition, language understanding, reasoning, planning, navigation, manipulation, and so on. Without a unifying framework, these attempts soon hit an insurmountable wall of complexity: too many moving parts, too many interactions, too many bugs for poor human software engineers to cope with. Knowledge engineers believe AI is just an engineering problem, but we have not yet reached the point where engineering can take us the rest of the way. In 1962, when Kennedy gave his famous moon-shot speech, going to the moon was an engineering problem. In 1662, it wasn't, and that's closer to where AI is today.

In industry, there's no sign that knowledge engineering will ever be able to compete with machine learning outside of a few niche areas. Why pay experts to slowly and painfully encode knowledge into a form computers can understand, when you can extract it from data at a fraction of the cost? What about all the things the experts don't know but you can discover from data? And when data is not available, the cost of knowledge engineering seldom exceeds the benefit. Imagine if farmers had to engineer each cornstalk in turn, instead of sowing the seeds and letting them grow: we would all starve.

Another prominent machine-learning skeptic is the linguist Noam Chomsky. Chomsky believes that language must be innate, because the examples of grammatical sentences children hear are not enough to learn a grammar. This only puts the burden of learning language on evolution, however; it does not argue against the Master Algorithm but only against it being something like the brain. Moreover, if a universal grammar exists (as Chomsky believes), elucidating it is a step toward elucidating the Master Algorithm. The only way this is not the case is if language has nothing in common with other cognitive abilities, which is implausible given its evolutionary recency.

In any case, if we formalize Chomsky's "poverty of the stimulus" argument, we find that it's demonstrably false. In 1969, J. J. Horning proved that probabilistic context-free grammars can be learned from positive examples only, and stronger results have followed. (Context-free grammars are the linguist's bread and butter, and the probabilistic version models how likely each rule is to be used.) Besides, language learning doesn't happen in a vacuum; children get all sorts of cues from their parents and the environment. If we're able to learn language from a few years' worth of examples, it's partly because of the similarity between its structure and the structure of the world. This common structure is what we're interested in, and we know from Horning and others that it suffices.

More generally, Chomsky is critical of all statistical learning. He has a list of things statistical learners can't do, but the list is fifty years out of date. Chomsky seems to equate machine learning with behaviorism, where animal behavior is reduced to associating responses with rewards. But machine learning is not behaviorism. Modern learning algorithms can learn rich internal representations, not just pairwise associations between stimuli.

In the end, the proof is in the pudding. Statistical language learners work, and hand-engineered language systems don't. The first eye-opener came in the 1970s, when DARPA, the Pentagon's research arm, organized the first large-scale speech recognition project. To everyone's surprise, a simple sequential learner of the type Chomsky derided handily beat a sophisticated knowledge-based system. Learners like it are now used in just about every speech recognizer, including Siri. Fred Jelinek, head of the speech group at IBM, famously quipped that "every time I fire a linguist, the recognizer's performance goes up." Stuck in the knowledge-engineering mire, computational linguistics had a near-death experience in the late 1980s. Since then, learning-based methods have swept the field, to the point where it's hard to find a paper devoid of learning in a computational linguistics conference. Statistical parsers analyze language with accuracy close to that of humans, where hand-coded ones lagged far behind. Machine translation, spelling correction, part-of-speech tagging, word sense disambiguation, question answering, dialogue, summarization: the best systems in these areas all use learning. Watson, the *Jeopardy!* computer champion, would not have been possible without it.

To this Chomsky might reply that engineering successes are not proof of scientific validity. On the other hand, if your buildings collapse and your engines don't run, perhaps something is wrong with your theory of physics. Chomsky thinks linguists should focus on "ideal" speaker-listeners, as defined by him, and this gives him license to ignore things like the need for statistics in language learning. Perhaps it's not surprising, then, that few experimentalists take his theories seriously any more.

Another potential source of objections to the Master Algorithm is the notion, popularized by the psychologist Jerry Fodor, that the mind is composed of a set of modules with only limited communication between them. For example, when you watch TV your "higher brain" knows that it's only light flickering on a flat surface, but your visual system still sees three-dimensional shapes. Even if we believe in the modularity of mind, however, that does not imply that different modules use different learning algorithms. The same algorithm operating on, say, visual and verbal information may suffice.

Critics like Minsky, Chomsky, and Fodor once had the upper hand, but thankfully their influence has waned. Nevertheless, we should keep their criticisms in mind as we set out on the road to the Master Algorithm for two reasons. The first is that knowledge engineers faced many of the same problems machine learners do, and even if they didn't succeed, they learned many valuable lessons. The second is that learning and knowledge are intertwined in surprisingly subtle ways, as we'll soon find out. Unfortunately, the two camps often talk past each other. They speak different languages: machine learning speaks probability, and knowledge engineering speaks logic. Later in the book we'll see what to do about this.

Swan bites robot

"No matter how smart your algorithm, there are some things it just can't learn." Outside of AI and cognitive science, the most common objections to machine learning are variants of this claim. Nassim Taleb hammered on it forcefully in his book *The Black Swan*. Some events are simply not predictable. If you've only ever seen white swans, you think the probability of ever seeing a black one is zero. The financial melt-down of 2008 was a "black swan."

It's true that some things are predictable and some aren't, and the first duty of the machine learner is to distinguish between them. But the goal of the Master Algorithm is to learn everything that *can* be known, and that's a vastly wider domain than Taleb and others imagine. The housing bust was far from a black swan; on the contrary, it was widely predicted. Most banks' models failed to see it coming, but that was due to well-understood limitations of those models, not limitations of machine learning in general. Learning algorithms are quite capable of accurately predicting rare, never-before-seen events; you could even say that that's what machine learning is all about. What's the probability of a black swan if you've never seen one? How about it's the fraction of known species that belatedly turned out to have black specimens? This is only a crude example; we'll see many deeper ones in this book.

A related, frequently heard objection is "Data can't replace human intuition." In fact, it's the other way around: human intuition can't replace data. Intuition is what you use when you don't know the facts, and since you often don't, intuition is precious. But when the evidence is before you, why would you deny it? Statistical analysis beats talent scouts in baseball (as Michael Lewis memorably documented in *Moneyball*), it beats connoisseurs at wine tasting, and every day we see new examples of what it can do. Because of the influx of data, the boundary between evidence and intuition is shifting rapidly, and as with any revolution, entrenched ways have to be overcome. If I'm the expert on X at company Y, I don't like to be overridden by some guy with data. There's a saying in industry: "Listen to your customers, not to the HiPPO," HiPPO being short for "highest paid person's opinion." If you want to be tomorrow's authority, ride the data, don't fight it.

OK, some say, machine learning can find statistical regularities in data, but it will never discover anything deep, like Newton's laws. It arguably hasn't yet, but I bet it will. Stories of falling apples notwithstanding, deep scientific truths are not low-hanging fruit. Science goes through three phases, which we can call the Brahe, Kepler, and Newton phases. In the Brahe phase, we gather lots of data, like Tycho Brahe patiently recording the positions of the planets night after night, year after year. In the Kepler phase, we fit empirical laws to the data, like Kepler did to the planets' motions. In the Newton phase, we discover the deeper truths. Most science consists of Brahe- and Kepler-like work; Newton moments are rare. Today, big data does the work of billions of Brahes, and machine learning the work of millions of Keplers. If—let's hope so—there are more Newton moments to be had, they are as likely to come from tomorrow's learning algorithms as from tomorrow's even more overwhelmed scientists, or at least from a combination of the two. (Of course, the Nobel prizes will go to the scientists, whether they have the key insights or just push the button. Learning algorithms have no ambitions of their own.) We'll see in this book what those algorithms might look like and speculate about what they might discover—such as a cure for cancer.

Is the Master Algorithm a fox or a hedgehog?

We need to consider one more potential objection to the Master Algorithm, perhaps the most serious one of all. It comes not from knowledge engineers or disgruntled experts, but from the machine-learning practitioners themselves. Putting that hat on for a moment, I might say: "But the Master Algorithm does not look like my daily life. I try hundreds of variations of many different learning algorithms on any given problem, and different algorithms do better on different problems. How could a single algorithm replace them all?"

To which the answer is: indeed. Wouldn't it be nice if, instead of trying hundreds of variations of many algorithms, we just had to try hundreds of variations of a single one? If we can figure out what's important and not so important in each one, what the important parts have in common and how they complement each other, we can, indeed, synthesize a Master Algorithm from them. That's what we're going to do in this book, or as close to it as we can. Perhaps you, dear reader, will have some ideas of your own as you read it.

How complex will the Master Algorithm be? Thousands of lines of code? Millions? We don't know yet, but machine learning has a delightful history of simple algorithms unexpectedly beating very fancy ones. In a famous passage of his book *The Sciences of the Artificial*, AI pioneer and Nobel laureate Herbert Simon asked us to consider an ant laboriously making its way home across a beach. The ant's path is complex, not because the ant itself is complex but because the environment is full of dunelets to climb and pebbles to get around. If we tried to model the ant by programming in every possible path, we'd be doomed. Similarly, in machine learning the complexity is in the data; all the Master Algorithm has to do is assimilate it, so we shouldn't be surprised if it turns out to be simple. The human hand is simple—four fingers, one opposable thumb—and yet it can make and use an infinite variety of tools. The Master Algorithm is to algorithms what the hand is to pens, swords, screwdrivers, and forks.

As Isaiah Berlin memorably noted, some thinkers are foxes—they know many small things—and some are hedgehogs—they know one big thing. The same is true of learning algorithms. I hope the Master Algorithm is a hedgehog, but even if it's a fox, we can't catch it soon enough. The biggest problem with today's learning algorithms is not that they are plural; it's that, useful as they are, they still don't do everything we'd like them to. Before we can discover deep truths with machine learning, we have to discover deep truths about machine learning.

What's at stake

Suppose you've been diagnosed with cancer, and the traditional treatments—surgery, chemotherapy, and radiation therapy—have failed. What happens next will determine whether you live or die. The first step is to get the tumor's genome sequenced. Companies like Foundation Medicine in Cambridge, Massachusetts, will do that for you: send them a sample of the tumor and they will send back a list of the known cancer-related mutations in its genome. This is needed because every cancer is different, and no single drug is likely to work for all. Cancers mutate as they spread through your body, and by natural selection, the mutations most resistant to the drugs you're taking are the most likely to grow. The right drug for you may be one that works for only 5 percent of patients, or you may need a combination of drugs that has never been tried before. Perhaps it will take a new drug designed specifically for your cancer, or a sequence of drugs to parry the cancer's adaptations. Yet these drugs may have side effects that are deadly for you but not most other people. No doctor can keep track of all the information needed to predict the best treatment for you, given your medical history and your cancer's genome. It's an ideal job for machine learning, and yet today's learners aren't up to it. Each has some of the needed capabilities but is missing others. The Master Algorithm is the complete package. Applying it to vast amounts of patient and drug data, combined with knowledge mined from the biomedical literature, is how we will cure cancer.

A universal learner is sorely needed in many other areas, from life-and-death to mundane situations. Picture the ideal recommender system, one that recommends the books, movies, and gadgets you would pick for yourself if you had the time to check them all out. Amazon's algorithm is a very far cry from it. That's partly because it doesn't have enough data—mainly it just knows which items you previously bought from Amazon—but if you went hog wild and gave it access to your complete stream of consciousness from birth, it wouldn't know what to do with it. How do you transmute the kaleidoscope of your life, the myriad different choices you've made, into a coherent picture of who you are and what you want? This is well beyond the ken of today's learners, but given enough data, the Master Algorithm should be able to understand you roughly as well as your best friend.

Someday there'll be a robot in every house, doing the dishes, making the beds, even looking after the children while the parents work. How soon depends on how hard finding the Master Algorithm turns out to be. If the best we can do is combine many different learners, each of which solves only a small part of the AI problem, we'll soon run into the complexity wall. This piecemeal approach worked for *Jeopardy!*, but few believe tomorrow's housebots will be Watson's grandchildren. It's not that the Master Algorithm will single-handedly crack AI; there'll still be great feats of engineering to perform, and Watson is a good preview of them. But the 80/20 rule applies: the Master Algorithm will be 80 percent of the solution and 20 percent of the work, so it's surely the best place to start.

The Master Algorithm's impact on technology will not be limited to AI. A universal learner is a phenomenal weapon against the complexity monster. Systems that today are too complex to build will no longer be. Computers will do more with less help from us. They will not repeat the same mistakes over and over again, but learn with practice, like people do. Sometimes, like the butlers of legend, they'll even guess what we want before we express it. If computers make us smarter, computers running the Master Algorithm will make us feel like geniuses. Technological progress will noticeably speed up, not just in computer science but in many different fields. This in turn will add to economic growth and speed poverty's decline. With the Master Algorithm to help synthesize and distribute knowledge, the intelligence of an organization will be more than the sum of its parts, not less. Routine jobs will be automated and replaced by more interesting ones. Every job will be done better than it is today, whether by a better-trained human, a computer, or a combination of the two. Stock-market crashes will be fewer and smaller. With a fine grid of sensors covering the globe and learned models to make sense of its output moment by moment, we will no longer be flying blind; the health of our planet will take a turn for the better. A model of you will negotiate the world on your behalf, playing elaborate games with other people's and entities' models. And as a result of all this, our lives will be longer, happier, and more productive.

Because the potential impact is so great, it would behoove us to try to invent the Master Algorithm even if the odds of success were low. And even if it takes a long time, searching for a universal learner has many immediate benefits. One is the better understanding of machine learning that a unified view enables. Too many business decisions are made with scant understanding of the analytics underpinning them, but it doesn't have to be that way. To use a technology, we don't need to master its inner workings, but we do need to have a good conceptual model of it. We need to know how to find a station on the radio, or change the volume. Today, those of us who aren't machine-learning experts have no conceptual model of what a learner does. The algorithms we drive when we use Google, Facebook, or the latest analytics suite are a bit like a black limo with tinted windows that mysteriously shows up at our door one night: Should we get in? Where will it take us? It's time to get in the driver's seat. Knowing the assumptions that different learners make will help us pick the right one for the job, instead of going with a random one that fell into our lap—and then suffering with it for years, painfully rediscovering what we should have known from the start. By knowing what learners optimize, we can make certain they optimize what we care about, rather than what came in the box. Perhaps most important, once we know how a particular learner arrives at its conclusions, we'll know what to make of that information—what to believe, what to return to the manufacturer, and how to get a better result next time around. And with the universal learner we'll develop in this book as the conceptual model, we can do all this without cognitive overload. Machine learning is simple at heart; we just need to peel away the layers of math and jargon to reveal the innermost Russian doll.

These benefits apply in both our personal and professional lives. How do I make the best of the trail of data that my every step in the modern world leaves? Every transaction works on two levels: what it accomplishes for you and what it teaches the system you just interacted with. Being aware of this is the first step to a happy life in the twenty-first century. Teach the learners, and they will serve you; but first you need to understand them. What in my job can be done by a learning algorithm, what can't, and—most important—how can I take advantage of machine learning to do it better? The computer is your tool, not your adversary. Armed with machine learning, a manager becomes a supermanager, a scientist a superscientist, an engineer a superengineer. The future belongs to those who understand at a very deep level how to combine their unique expertise with what algorithms do best.

But perhaps the Master Algorithm is a Pandora's box best left closed. Will computers enslave us or even exterminate us? Will machine learning be the handmaiden of dictators or evil corporations? Knowing where machine learning is headed will help us to understand what to worry about, what not, and what to do about it. The *Terminator* scenario, where a super-AI becomes sentient and subdues mankind with a robot army, has no chance of coming to pass with the kinds of learning algorithms we'll meet in this book. Just because computers can learn doesn't mean they magically acquire a will of their own. Learners learn to achieve the goals we set them; they don't get to change the goals. Rather, we need to worry about them trying to serve us in ways that do more harm than good because they don't know any better, and the cure for that is to teach them better.

Most of all, we have to worry about what the Master Algorithm could do in the wrong hands. The first line of defense is to make sure the good guys get it first—or, if it's not clear who the good guys are, to make sure it's open-sourced. The second is to realize that, no matter how good the learning algorithm is, it's only as good as the data it gets. He who controls the data controls the learner. Your reaction to the datafication of life should not be to retreat to a log cabin—the woods, too, are full of sensors—but to aggressively seek control of the data that matters to you. It's good to have recommenders that find what you want and bring it to you; you'd feel lost without them. But they should bring you what *you* want, not what someone else wants you to have. Control of data and ownership of the models learned from it is what many of the twenty-first century's battles will be about—between governments, corporations, unions, and individuals. But you also have an ethical duty to share data for the common good. Machine learning alone will not cure cancer; cancer patients will, by sharing their data for the benefit of future patients.

A different theory of everything

Science today is thoroughly balkanized, a Tower of Babel where each subcommunity speaks its own jargon and can see only into a few adjacent subcommunities. The Master Algorithm would provide a unifying view of all of science and potentially lead to a new theory of everything. At first this may seem like an odd claim. What machine learning does is induce theories from data. How could the Master Algorithm itself grow into a theory? Isn't string theory the theory of everything, and the Master Algorithm nothing like it?

To answer these questions, we have to first understand what a scientific theory is and is not. A theory is a set of constraints on what the world could be, not a complete description of it. To obtain the latter, you have to combine the theory with data. For example, consider Newton's second law. It says that force equals mass times acceleration, or $F = ma$. It does not say what the mass or acceleration of any object are, or the forces acting on it. It only requires that, if the mass of an object is m and its acceleration is a, then the total force on it must be ma. It removes some of the universe's degrees of freedom, but not all. The same is true of all other physical theories, including relativity, quantum mechanics, and string theory, which are, in effect, refinements of Newton's laws.

The power of a theory lies in how much it simplifies our description of the world. Armed with Newton's laws, we only need to know the masses, positions, and velocities of all objects at one point in time; their positions and velocities at all times follow. So Newton's laws reduce our description of the world by a factor of the number of distinguishable instants in the history of the universe, past and future. Pretty amazing! Of course, Newton's laws are only an approximation of the true laws of physics, so let's replace them with string theory, ignoring all its problems and the question of whether it can ever be empirically validated. Can we do better? Yes, for two reasons.

The first is that, in reality, we never have enough data to completely determine the world. Even ignoring the uncertainty principle, precisely knowing the positions and velocities of all particles in the world at some point in time is not remotely feasible. And because the laws of physics are chaotic, uncertainty compounds over time, and pretty soon they determine very little indeed. To accurately describe the world, we need a fresh batch of data at regular intervals. In effect, the laws of physics only tell us what happens locally. This drastically reduces their power.

The second problem is that, even if we had complete knowledge of the world at some point in time, the laws of physics would still not allow us to determine its past and future. This is because the sheer amount of computation required to make those predictions would be beyond the capabilities of any imaginable computer. In effect, to perfectly simulate the universe we would need another, identical universe. This is why string theory is mostly irrelevant outside of physics. The theories we have in biology, psychology, sociology, or economics are not corollaries of the laws of physics; they had to be created from scratch. We assume that they are approximations of what the laws of physics would predict when applied at the scale of cells, brains, and societies, but there's no way to know.

Unlike the theories of a given field, which only have power within that field, the Master Algorithm has power across all fields. Within field X, it has less power than field X's prevailing theory, but across all fields—when we consider the whole world—it has vastly more power than any other theory. The Master Algorithm is the germ of every theory; all we need to add to it to obtain theory X is the minimum amount of data required to induce it. (In the case of physics, that would be just the results of perhaps a few hundred key experiments.) The upshot is that, pound for pound, the Master Algorithm may well be the best starting point for a theory of everything we'll ever have. *Pace* Stephen Hawking, it may ultimately tell us more about the mind of God than string theory.

Some may say that seeking a universal learner is the epitome of techno-hubris. But dreaming is not hubris. Maybe the Master Algorithm will take its place among the great chimeras, alongside the philosopher's stone and the perpetual motion machine. Or perhaps it will be more like finding the longitude at sea, given up as too difficult until a lone genius solved it. More likely, it will be the work of generations, raised stone by stone like a cathedral. The only way to find out is to get up early one day and set out on the journey.

Candidates that don't make the cut

So, if the Master Algorithm exists, what is it? A seemingly obvious candidate is memorization: just remember everything you've seen; after a while you'll have seen everything there is to see, and therefore know everything there is to know. The problem with this is that, as Heraclitus said, you never step in the same river twice. There's far more to see than you ever could. No matter how many snowflakes you've examined, the next one will be different. Even if you had been present at the Big Bang and everywhere since, you would still have seen only a tiny fraction of what you could see in the future. If you had witnessed life on Earth up to ten thousand years ago, that would not have prepared you for what was to come. Someone who grew up in one city doesn't become paralyzed when they move to another, but a robot capable only of memorization would. Besides, knowledge is not just a long list of facts. Knowledge is general, and has structure. "All humans are mortal" is much more succinct than seven billion statements of mortality, one for each human. Memorization gives us none of these things.

Another candidate Master Algorithm is the microprocessor. After all, the one in your computer can be viewed as a single algorithm whose job is to execute other algorithms, like a universal Turing machine; and it can run any imaginable algorithm, up to its limits of memory and speed. In effect, to a microprocessor an algorithm is just another kind of data. The problem here is that, by itself, the microprocessor doesn't know how to do anything; it just sits there idle all day. Where do the algorithms it runs come from? If they were coded up by a human programmer, no learning is involved. Nevertheless, there's a sense in which the microprocessor is a good analog for the Master Algorithm. A microprocessor is not the best hardware for running any particular algorithm. That would be an ASIC (application-specific integrated circuit) designed very precisely for that algorithm. Yet microprocessors are what we use for almost all applications, because their flexibility trumps their relative inefficiency. If we had to build an ASIC for every new application, the Information Revolution would never have happened. Similarly, the Master Algorithm is not the best algorithm for learning any particular piece of knowledge; that would be an algorithm that already encodes most of that knowledge (or all of it, making the data superfluous). The point, however, is to induce the knowledge from data, because it's easier and costs less; so the more general the learning algorithm, the better.

An even more extreme candidate is the humble NOR gate: a logic switch whose output is 1 only if its inputs are both 0. Recall that all computers are made of logic gates built out of transistors, and all computations can be reduced to combinations of AND, OR, and NOT gates. A NOR gate is just an OR gate followed by a NOT gate: the negation of a disjunction, as in "I'm happy as long as I'm not starving or sick." AND, OR and NOT can all be implemented using NOR gates, so NOR can do everything, and in fact it's all some microprocessors use. So why can't it be the Master Algorithm? It's certainly unbeatable for simplicity. Unfortunately, a NOR gate is not the Master Algorithm any more than a Lego brick is the universal toy. It can certainly be a universal building block for toys, but a pile of Legos doesn't spontaneously assemble itself into a toy. The same applies to other simple computation schemes, like Petri nets or cellular automata.

Moving on to more sophisticated alternatives, what about the queries that any good database engine can answer, or the simple algorithms in a statistical package? Aren't those enough? These are bigger Lego bricks, but they're still only bricks. A database engine never discovers anything new; it just tells you what it knows. Even if all the humans in a database are mortal, it doesn't occur to it to generalize mortality to other humans. (Database engineers would blanch at the thought.) Much of statistics is about testing hypotheses, but someone has to formulate them in the first place. Statistical packages can do linear regression and other simple procedures, but these have a very low limit on what they can learn, no matter how much data you feed them. The better packages cross into the gray zone between statistics and machine learning, but there are still many kinds of knowledge they can't discover.

OK, it's time to come clean: the Master Algorithm is the equation $U(X) = 0$. Not only does it fit on a T-shirt; it fits on a postage stamp. Huh? $U(X) = 0$ just says that some (possibly very complex) function U of some (possibly very complex) variable X is equal to 0. Every equation can be reduced to this form; for example, $F = ma$ is equivalent to $F - ma = 0$, so if you think of $F - ma$ as a function U of F, voilà: $U(F) = 0$. In general, X could be any input and U could be any algorithm, so surely the Master Algorithm can't be any more general than this; and since we're looking for the most general algorithm we can find, this must be it. I'm just kidding, of course, but this particular failed candidate points to a real danger in machine learning: coming up with a learner that's so general, it doesn't have enough content to be useful.

So what's the least content a learner can have in order to be useful? How about the laws of physics? After all, everything in the world obeys them (we believe), and they gave rise to evolution and (through it) the brain. Well, perhaps the Master Algorithm is implicit in the laws of physics, but if so, then we need to make it explicit. Just throwing data at the laws of physics won't result in any new laws. Here's one way to think about it: perhaps some field's master theory is just the laws of physics compiled into a more convenient form for that field, but if so then we need an algorithm that finds a shortcut from that field's data to its theory, and it's not clear the laws of physics can be of any help with this. Another issue is that, if the laws of physics were different, the Master Algorithm would presumably still be able to discover them in many cases. Mathematicians like to say that God can disobey the laws of physics, but even he cannot defy the laws of logic. This may be so, but the laws of logic are for deduction; what we need is something equivalent, but for induction.

The five tribes of machine learning

Of course, we don't have to start from scratch in our hunt for the Master Algorithm. We have a few decades of machine learning research to draw on. Some of the smartest people on the planet have devoted their lives to inventing learning algorithms, and some would even claim that they already have a universal learner in hand. We will stand on the shoulders of these giants, but take such claims with a grain of salt. Which raises the question: how will we know when we've found the Master Algorithm? When the same learner, with only parameter changes and minimal input aside from the data, can understand video and text as well as humans, and make significant new discoveries in biology, sociology, and other sciences. Clearly, by this standard no learner has yet been demonstrated to be the Master Algorithm, even in the unlikely case one already exists.

Crucially, the Master Algorithm is not required to start from scratch in each new problem. That bar is probably too high for *any* learner to meet, and it's certainly very unlike what people do. For example, language does not exist in a vacuum; we couldn't understand a sentence without our knowledge of the world it refers to. Thus, when learning to read, the Master Algorithm can rely on having previously learned to see, hear, and control a robot. Likewise, a scientist does not just blindly fit models to data; he can bring all his knowledge of the field to bear on the problem. Therefore, when making discoveries in biology, the Master Algorithm can first read all the biology it wants, relying on having previously learned to read. The Master Algorithm is not just a passive consumer of data; it can interact with its environment and actively seek the data it wants, like Adam, the robot scientist, or like any child exploring her world.

Our search for the Master Algorithm is complicated, but also enlivened, by the rival schools of thought that exist within machine learning. The main ones are the symbolists, connectionists, evolutionaries, Bayesians, and analogizers. Each tribe has a set of core beliefs, and a particular problem that it cares most about. It has found a solution to that problem, based on ideas from its allied fields of science, and it has a master algorithm that embodies it.

For symbolists, all intelligence can be reduced to manipulating symbols, in the same way that a mathematician solves equations by replacing expressions by other expressions. Symbolists understand that you can't learn from scratch: you need some initial knowledge to go with the data. They've figured out how to incorporate preexisting knowledge into learning, and how to combine different pieces of knowledge on the fly in order to solve new problems. Their master algorithm is inverse deduction, which figures out what knowledge is missing in order to make a deduction go through, and then makes it as general as possible.

For connectionists, learning is what the brain does, and so what we need to do is reverse engineer it. The brain learns by adjusting the strengths of connections between neurons, and the crucial problem is figuring out which connections are to blame for which errors and changing them accordingly. The connectionists' master algorithm is backpropagation, which compares a system's output with the desired one and then successively changes the connections in layer after layer of neurons so as to bring the output closer to what it should be.

Evolutionaries believe that the mother of all learning is natural selection. If it made us, it can make anything, and all we need to do is simulate it on the computer. The key problem that evolutionaries solve is learning structure: not just adjusting parameters, like backpropagation does, but creating the brain that those adjustments can then fine-tune. The evolutionaries' master algorithm is genetic programming, which mates and evolves computer programs in the same way that nature mates and evolves organisms.

Bayesians are concerned above all with uncertainty. All learned knowledge is uncertain, and learning itself is a form of uncertain inference. The problem then becomes how to deal with noisy, incomplete, and even contradictory information without falling apart. The solution is probabilistic inference, and the master algorithm is Bayes' theorem and its derivates. Bayes' theorem tells us how to incorporate new evidence into our beliefs, and probabilistic inference algorithms do that as efficiently as possible.

For analogizers, the key to learning is recognizing similarities between situations and thereby inferring other similarities. If two patients have similar symptoms, perhaps they have the same disease. The key problem is judging how similar two things are. The analogizers' master algorithm is the support vector machine, which figures out which experiences to remember and how to combine them to make new predictions.

Each tribe's solution to its central problem is a brilliant, hard-won advance. But the true Master Algorithm must solve all five problems, not just one. For example, to cure cancer we need to understand the metabolic networks in the cell: which genes regulate which others, which chemical reactions the resulting proteins control, and how adding a new molecule to the mix would affect the network. It would be silly to try to learn all of this from scratch, ignoring all the knowledge that biologists have painstakingly accumulated over the decades. Symbolists know how to combine this knowledge with data from DNA sequencers, gene expression microarrays, and so on, to produce results that you couldn't get with either alone. But the knowledge we obtain by inverse deduction is purely qualitative; we need to learn not just who interacts with whom, but how much, and backpropagation can do that. Nevertheless, both inverse deduction and backpropagation would be lost in space without some basic structure on which to hang the interactions and parameters they find, and genetic programming can discover it. At this point, if we had complete knowledge of the metabolism and all the data relevant to a given patient, we could figure out a treatment for her. But in reality the information we have is always very incomplete, and even incorrect in places; we need to make headway despite that, and that's what probabilistic inference is for. In the hardest cases, the patient's cancer looks very different from previous ones, and all our learned knowledge fails. Similarity-based algorithms can save the day by seeing analogies between superficially very different situations, zeroing in on their essential similarities and ignoring the rest.

In this book we will synthesize a single algorithm with all these capabilities:

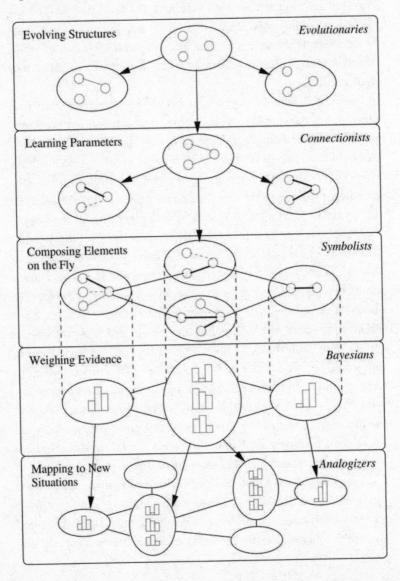

Our quest will take us across the territory of each of the five tribes. The border crossings, where they meet, negotiate and skirmish, will be the trickiest part of the journey. Each tribe has a different piece of the puzzle, which we must gather. Machine learners, like all scientists, resemble the blind men and the elephant: one feels the trunk and thinks it's a snake, another leans against the leg and thinks it's a tree, yet another touches the tusk and thinks it's a bull. Our aim is to touch each part without jumping to conclusions; and once we've touched all of them, we will try to picture the whole elephant. It's far from obvious how to combine all the pieces into one solution—impossible, according to some— but this is what we will do.

The algorithm we'll arrive at is not yet the Master Algorithm, for reasons we'll see, but it's the closest anyone has come. And we'll gather enough riches along the way to make Croesus envious. Nevertheless, this book is only part one of the Master Algorithm saga. Part two's protagonist is you, dear reader. Your mission, should you choose to accept it, is to go the rest of the way and bring back the prize. I will be your humble guide in part one, from here to the edge of the known world. Do I hear you protest that you don't know enough, or algorithms are not your forte? Fear not. Computer science is still young, and unlike in physics or biology, you don't need a PhD to start a revolution. (Just ask Bill Gates, Messrs. Sergey Brin and Larry Page, or Mark Zuckerberg.) Insight and persistence are what counts.

Are you ready? Our journey begins with a visit to the symbolists, the tribe with the oldest roots.

CHAPTER THREE

Hume's Problem of Induction

Are you a rationalist or an empiricist?

Rationalists believe that the senses deceive and that logical reasoning is the only sure path to knowledge. Empiricists believe that all reasoning is fallible and that knowledge must come from observation and experimentation. The French are rationalists; the Anglo-Saxons (as the French call them) are empiricists. Pundits, lawyers, and mathematicians are rationalists; journalists, doctors, and scientists are empiricists. *Murder, She Wrote* is a rationalist TV crime show; *CSI: Crime Scene Investigation* is an empiricist one. In computer science, theorists and knowledge engineers are rationalists; hackers and machine learners are empiricists.

The rationalist likes to plan everything in advance before making the first move. The empiricist prefers to try things and see how they turn out. I don't know if there's a gene for rationalism or one for empiricism, but looking at my computer scientist colleagues, I've observed time and again that they are almost like personality traits: some people are rationalistic to the core and could never have been otherwise; and others are empiricist through and through, and that's what they'll always be. The two sides can converse with each other and sometimes draw on each other's results, but they can understand each other only so much. Deep down each believes that what the other does is secondary, and not very interesting.

Rationalists and empiricists have probably been around since the dawn of *Homo sapiens*. Before setting out on a hunt, Caveman Bob spent a long time sitting in his cave figuring out where the game would be. In the meantime, Cavewoman Alice was out systematically surveying the territory. Since both kinds are still with us, it's probably safe to say that neither approach was better. You might think that machine learning is the final triumph of the empiricists, but the truth is more subtle, as we'll soon see.

Rationalism versus empiricism is a favorite question of philosophers. Plato was an early rationalist, and Aristotle an early empiricist. But the debate really took off during the Enlightenment, with a trio of great thinkers on each side: Descartes, Spinoza, and Leibniz were the leading rationalists; Locke, Berkeley, and Hume were their empiricist counterparts. Trusting in their powers of reasoning, the rationalists concocted theories of the universe that—to put it gently—did not stand the test of time, but they also invented fundamental mathematical techniques like calculus and analytical geometry. The empiricists were altogether more practical, and their influence is everywhere from the scientific method to the Constitution of the United States.

David Hume was the greatest of the empiricists and the greatest English-speaking philosopher of all time. Thinkers like Adam Smith and Charles Darwin count him among their key influences. You could also say he's the patron saint of the symbolists. He was born in Scotland in 1711 and spent most of his life in eighteenth-century Edinburgh, a prosperous city full of intellectual ferment. A man of genial disposition, he was nevertheless an exacting skeptic who spent much of his time debunking the myths of his age. He also took the empiricist train of thought that Locke had started to its logical conclusion and asked a question that has since hung like a sword of Damocles over all knowledge, from the most trivial to the most advanced: How can we ever be justified in generalizing from what we've seen to what we haven't? Every learning algorithm is, in a sense, an attempt to answer this question.

Hume's question is also the departure point for our journey. We'll start by illustrating it with an example from daily life and meeting its modern embodiment in the famous "no free lunch" theorem. Then we'll see the symbolists' answer to Hume. This leads us to the most important problem in machine learning: overfitting, or hallucinating patterns that aren't really there. We'll see how the symbolists solve it, and how machine learning is at heart a kind of alchemy, transmuting data into knowledge with the aid of a philosopher's stone. For the symbolists, the philosopher's stone is knowledge itself. In the next four chapters we'll study the solutions of the other tribes' alchemists.

To date or not to date?

You have a friend you really like, and you want to ask her out on a date. You have a hard time dealing with rejection, though, and you're only going to ask her if you're pretty sure she'll say yes. It's Friday evening, and there you sit with cell phone in hand, trying to decide whether or not to call her. You remember that the previous time you asked her, she said no. But why? Two times before that she said yes, and the one before those she said no. Maybe there are days she doesn't like to go out? Or maybe she likes clubbing but not dinner dates? Being of an unusually systematic nature, you put down the phone and jot down what you can remember about those previous occasions:

Occasion	Day of Week	Type of Date	Weather	TV Tonight	Date?
1	Weekday	Dinner	Warm	Bad	No
2	Weekend	Club	Warm	Bad	Yes
3	Weekend	Club	Warm	Bad	Yes
4	Weekend	Club	Cold	Good	No
Today	Weekend	Club	Cold	Bad	?

So … what shall it be? Date or no date? Is there a pattern that distinguishes the yeses from the nos? And, most important, what does that pattern say about today?

Clearly, there's no single factor that correctly predicts the answer: on some weekends she likes to go out, and on some she doesn't; sometimes she likes to go clubbing, and sometimes she doesn't, and so on. What about a combination of factors? Maybe she likes to go clubbing on weekends? No, occasion number 4 crosses that one out. Or maybe she only likes to go out on warm weekend nights? Bingo! That works! In which case, looking at the frosty weather outside, tonight doesn't look promising. But wait! What if she likes to go clubbing when there's nothing good on TV? That also works, and that means today is a yes! Quick, call her before it gets too late. But wait a second. How do you know this is the right pattern? You've found two that agree with your previous experience, but they make opposite predictions. Come to think of it, what if she only goes clubbing when the weather is nice? Or she goes out on weekends when there's nothing to watch on TV? Or—

At this point you crumple your notes in frustration and fling them into the wastebasket. There's no way to know! What can you do? The ghost of Hume nods sadly over your shoulder. *You have no basis to pick one generalization over another.* Yes and no are equally legitimate answers to the question "What will she say?" And the clock is ticking. Bitterly, you fish out a quarter from your pocket and prepare to flip it.

You're not the only one in dire straits—so are we. We've only just set out on our road to the Master Algorithm and already we seem to have run into an insurmountable obstacle. Is there *any* way to learn something from the past that we can be confident will apply in the future? And if there isn't, isn't machine learning a hopeless enterprise? For that matter, isn't all of science, even all of human knowledge, on rather shaky ground?

It's not like big data would solve the problem. You could be super-Casanova and have dated millions of women thousands of times each, but your master database still wouldn't answer the question of what *this* woman is going to say *this* time. Even if today is exactly like some previous occasion when she said yes—same day of week, same type of date, same weather, and same shows on TV—that still doesn't mean that this time she will say yes. For all you know, her answer is determined by some factor that you didn't think of or don't have access to. Or maybe there's no rhyme or reason to her answers: they're random, and you're just spinning your wheels trying to find a pattern in them.

Philosophers have debated Hume's problem of induction ever since he posed it, but no one has come up with a satisfactory answer. Bertrand Russell liked to illustrate the problem with the story of the inductivist turkey. On his first morning at the farm, the turkey was fed at 9:00 a.m., but being a good inductivist, he didn't jump to conclusions. He first collected many observations on many different days under many different circumstances. Having been fed consistently at 9:00 a.m. for many consecutive days, he finally concluded that yes, he would always be fed at 9:00 a.m. Then came the morning of Christmas eve, and his throat was cut.

It would be nice if Hume's problem was just a cute philosophical conundrum we could ignore, but we can't. For example, Google's business is based on guessing which web pages you're looking for when you type some keywords into the search box. Their key asset is massive logs of search queries people have entered in the past and the links they clicked on in the corresponding results pages. But what do you do if someone types in a combination of keywords that's not in the log? And even if it is, how can you be confident that the current user wants the same pages as the previous ones?

How about we just *assume* that the future will be like the past? This is certainly a risky assumption. (It didn't work for the inductivist turkey.) On the other hand, without it all knowledge is impossible, and so is life. We'd rather stay alive, even if precariously. Unfortunately, even with that assumption we're not out of the woods. It takes care of the "trivial" cases: If I'm a doctor and patient B has exactly the same symptoms as patient A, I assume that the diagnosis is the same. But if patient B's symptoms don't exactly match anyone else's, I'm still in the dark.

This is the machine-learning problem: generalizing to cases that *we haven't seen before*.

But perhaps that's not such a big deal? With enough data, won't most cases be in the "trivial" category? No. We saw in the previous chapter why memorization won't work as a universal learner, but now we can make it more quantitative. Suppose you have a database with a trillion records, each with a thousand Boolean fields (i.e., each field is the answer to a yes/no question). That's pretty big. What fraction of the possible cases have you seen? (Take a guess before you read on.) Well, the number of possible answers is two for each question, so for two questions it's two times two (yes-yes, yes-no, no-yes, and no-no), for three questions it's two cubed ($2 \times 2 \times 2 = 2^3$), and for a thousand questions it's two raised to the power of a thousand (2^{1000}). The trillion records in your database are one-gazillionth of 1 percent of 2^{1000}, where "gazillionth" means "zero point 286 zeros followed by 1." Bottom line: no matter how much data you have—tera- or peta- or exa- or zetta- or yottabytes—you've basically seen *nothing*. The chances that the new case you need to make a decision on is already in the database are so vanishingly small that, without generalization, you won't even get off the ground.

If this all sounds a bit abstract, suppose you're a major e-mail provider, and you need to label each incoming e-mail as spam or not spam. You may have a database of a trillion past e-mails, each already labeled as spam or not, but that won't save you, since the chances that every new e-mail will be an exact copy of a previous one are just about zero. You have no choice but to try to figure out at a more general level what distinguishes spam from nonspam. And, according to Hume, there's no way to do that.

The "no free lunch" theorem

Two hundred and fifty years after Hume set off his bombshell, it was given elegant mathematical form by David Wolpert, a physicist turned machine learner. His result, known as the "no free lunch" theorem, sets a limit on how good a learner can be. The limit is pretty low: no learner can be better than random guessing! OK, we can go home: the Master Algorithm is just flipping coins. Seriously, though, how is it that no learner can beat coin flipping? And if that's so, how come the world is full of highly successful learners, from spam filters to (any day now) self-driving cars?

The "no free lunch" theorem is a lot like the reason Pascal's wager fails. In his *Pensées*, published in 1669, Pascal said we should believe in the Christian God because if he exists that gains us eternal life, and if he doesn't we lose very little. This was a remarkably sophisticated argument for the time, but as Diderot pointed out, an imam could make the same argument for believing in Allah. And if you pick the wrong god, the price you pay is eternal hell. On balance, considering the wide variety of possible gods, you're no better off picking a particular one to believe in than you are picking any other. For every god that says "do this," there's another that says "no, do that." You may as well just forget about god and enjoy life without religious constraints.

Replace "god" with "learning algorithm" and "eternal life" with "accurate prediction," and you have the "no free lunch" theorem. Pick your favorite learner. (We'll see many in this book.) For every world where it does better than random guessing, I, the devil's advocate, will deviously construct one where it does worse by the same amount. All I have to do is flip the labels of all *unseen* instances. Since the labels of the observed ones agree, there's no way your learner can distinguish between the world and the antiworld. On average over the two, it's as good as random guessing. And therefore, on average over all possible worlds, pairing each world with its antiworld, your learner is equivalent to flipping coins.

Don't give up on machine learning or the Master Algorithm just yet, though. We don't care about all possible worlds, only the one we live in. If we know something about the world and incorporate it into our learner, it now has an advantage over random guessing. To this Hume would reply that that knowledge must itself have come from induction and is therefore fallible. That's true, even if the knowledge was encoded into our brains by evolution, but it's a risk we'll have to take. We can also ask whether there's a nugget of knowledge so incontestable, so fundamental, that we can build all induction on top of it. (Something like Descartes' "I think, therefore I am," although it's hard to see how to turn that one into a learning algorithm.) I think the answer is yes, and we'll see what that nugget is in Chapter 9.

In the meantime, the practical consequence of the "no free lunch" theorem is that there's no such thing as learning without knowledge. Data alone is not enough. Starting from scratch will only get you to scratch. Machine learning is a kind of knowledge pump: we can use it to extract a lot of knowledge from data, but first we have to prime the pump.

Machine learning is what mathematicians call an ill-posed problem: it doesn't have a unique solution. Here's a simple ill-posed problem: Which two numbers add up to 1,000? Assuming the numbers are positive, there are five hundred possible answers: 1 and 999, 2 and 998, and so on. The only way to solve an ill-posed problem is to introduce additional assumptions. If I tell you the second number is triple the first, bingo: the answer is 250 and 750.

Tom Mitchell, a leading symbolist, calls it "the futility of bias-free learning." In ordinary life, *bias* is a pejorative word: preconceived notions are bad. But in machine learning, preconceived notions are indispensable; you can't learn without them. In fact, preconceived notions are also indispensable to human cognition, but they're hardwired into the brain, and we take them for granted. It's biases over and beyond those that are questionable.

Aristotle said that there is nothing in the intellect that was not first in the senses. Leibniz added, "Except the intellect itself." The human brain is not a blank slate because it's not a slate. A slate is passive, something you write on, but the brain actively processes the information it receives. Memory is the slate it writes on, and it does start out blank. On the other hand, a computer *is* a blank slate until you program it; the active process itself has to be written into memory before anything can happen. Our goal is to figure out the simplest program we can write such that it will continue to write itself by reading data, without limit, until it knows everything there is to know.

Machine learning has an unavoidable element of gambling. In the first *Dirty Harry* movie, Clint Eastwood chases a bank robber, repeatedly firing at him. Finally, the robber is lying next to a loaded gun, unsure whether to spring for it. Did Harry fire six shots or only five? Harry sympathizes (so to speak): "You've got to ask yourself one question: 'Do I feel lucky?' Well, do you, punk?" That's the question machine learners have to ask themselves every day when they go to work: Do I feel lucky today? Just like evolution, machine learning doesn't get it right every time; in fact, errors are the rule, not the exception. But it's OK, because we discard the misses and build on the hits, and the cumulative result is what matters. Once we acquire a new piece of knowledge, it becomes a basis for inducing yet more knowledge. The only question is where to begin.

Priming the knowledge pump

In the *Principia*, along with his three laws of motion, Newton enunciates four rules of induction. Although these are much less well known than the physical laws, they are arguably as important. The key rule is the third one, which we can paraphrase thus:

> *Newton's Principle: Whatever is true of everything we've seen is true of everything in the universe.*

It's not an exaggeration to say that this innocuous-sounding statement is at the heart of the Newtonian revolution and of modern science. Kepler's laws applied to exactly six entities: the planets of the solar system known in his time. Newton's laws apply to every last speck of matter in the universe. The leap in generality between the two is staggering, and it's a direct consequence of Newton's principle. This one principle is all by itself a knowledge pump of phenomenal power. Without it there would be no laws of nature, only a forever incomplete patchwork of small regularities.

Newton's principle is the first unwritten rule of machine learning. We induce the most widely applicable rules we can and reduce their scope only when the data forces us to. At first sight this may seem ridiculously overconfident, but it's been working for science for over three hundred years. It's certainly possible to imagine a universe so varied and capricious that Newton's principle would systematically fail, but that's not our universe.

Newton's principle is only the first step, however. We still need to figure out what is true of everything we've seen—how to extract the regularities from the raw data. The standard solution is to assume we know the *form* of the truth, and the learner's job is to flesh it out. For example, in the dating problem you could assume that your friend's answer is determined by a single factor, in which case learning just consists of checking each known factor (day of week, type of date, weather, and TV programming) to see if it correctly predicts her answer every time. The problem, of course, is that none of them do! You gambled and failed. So you relax your assumptions a bit. What if your friend's answer is determined by a conjunction of two factors? With four factors, each with two possible values, there are twenty-four possibilities to check (six pairs of factors to pick from times two choices for each factor's value). Now we have an embarrassment of riches: four conjunctions of two factors correctly predict the outcome! What to do? If you're feeling lucky, you can just pick one of them and hope for the best. A more sensible option, though, is democracy: let them vote, and pick the winning prediction.

If all conjunctions of two factors fail, you can try all conjunctions of any number of factors. Machine learners and psychologists call these "conjunctive concepts." Dictionary definitions are conjunctive concepts: a chair has a seat *and* a back *and* some number of legs. Remove any of these and it's no longer a chair. A conjunctive concept is what Tolstoy had in mind when he wrote the opening sentence of *Anna Karenina*: "All happy families are alike; each unhappy family is unhappy in its own way." The same is true of individuals. To be happy, you need health, love, friends, money, a job you like, and so on. Take any of these away, and misery ensues.

In machine learning, examples of a concept are called positive examples, and counterexamples are called negative examples. If you're trying to learn to recognize cats in images, images of cats are positive examples and images of dogs are negative ones. If you compiled a database of families from the world's literature, the Karenins would be a negative example of a happy family, and there would be precious few positive examples.

Starting with restrictive assumptions and gradually relaxing them if they fail to explain the data is typical of machine learning, and the process is usually carried out automatically by the learner, without any help from you. First, it tries all single factors, then all conjunctions of two factors, then all conjunctions of three, and so on. But now we run into a problem: there are *a lot* of conjunctive concepts and not enough time to try them all out.

The dating example is a little deceptive because it's very small (four variables and four examples). But suppose now that you run an online dating service and you need to figure out which couples to match. If each user of your system has filled out a questionnaire with answers to fifty yes/no questions, each potential match is characterized by one hundred attributes, fifty from each member of the prospective couple. Based on the couples that have gone on a date and reported the outcome, can you find a conjunctive definition for the concept of a "good match"? There are 3^{100} possible definitions to try. (The three options for each attribute are yes, no, and not part of the concept.) Even with the fastest computer in the world, the couples will all be long gone—and your company bankrupt—by the time you're done, unless you're lucky and a very short definition hits the jackpot. So many rules, so little time. We need to do something smarter.

Here's one way. Suspend your disbelief and start by assuming that all matches are good. Then try excluding all matches that don't have some attribute. Repeat this for each attribute, and choose the one that excludes the most bad matches and the fewest good ones. Your definition now looks something like, say, "It's a good match only if he's outgoing." Now try adding every other attribute to that in turn, and choose the one that excludes the most remaining bad matches and fewest remaining good ones. Perhaps the definition is now "It's a good match only if he's outgoing and so is she." Try adding a third attribute to those two, and so on. Once you've excluded all the bad matches, you're done: you have a definition of the concept that includes all the positive examples and excludes all the negative ones. For example: "A couple is a good match only if they're both outgoing, he's a dog person, and she's not a cat person." You can now throw away the data and keep only this definition, since it encapsulates all that's relevant for your purposes. This algorithm is guaranteed to finish in a reasonable amount of time, and it's also the first actual learner we meet in this book!

How to rule the world

Conjunctive concepts don't get you very far, though. The problem is that, as Rudyard Kipling said, "There are nine and sixty ways of constructing tribal lays, and every one of them is right." Real concepts are disjunctive. Chairs can have four legs or one, and sometimes none. You can win at chess in countless different ways. E-mails containing the word *Viagra* are probably spam, but so are e-mails containing "FREE!!!" Besides, all rules have exceptions. Some families manage to be dysfunctional yet happy. Birds fly, unless they're penguins, ostriches, cassowaries, or kiwis (or they've broken a wing, or are locked in a cage, or . . .).

What we need is to learn concepts that are defined by a set of rules, not just a single rule, such as:

> *If you liked* Star Wars, *episodes IV–VI, you'll like* Avatar.
> *If you liked* Star Trek: The Next Generation *and* Titanic, *you'll like* Avatar.
> *If you're a member of the Sierra Club and read science-fiction books, you'll like* Avatar.

Or:

If your credit card was used in China, Canada, and Nigeria yesterday,
it was stolen.

If your credit card was used twice after 11:00 p.m. on a weekday, it was
stolen.

If your credit card was used to purchase one dollar of gas, it was stolen.

(If you're wondering about the last rule, credit-card thieves used to routinely buy one dollar of gas to check that a stolen credit card was good before data miners caught on to the tactic.)

We can learn sets of rules like this one rule at a time, using the algorithm we saw before for learning conjunctive concepts. After we learn each rule, we discard the positive examples that it accounts for, so the next rule tries to account for as many of the remaining positive examples as possible, and so on until all are accounted for. It's an example of "divide and conquer," the oldest strategy in the scientist's playbook. We can also improve the algorithm for finding a single rule by keeping some number n of hypotheses around, not just one, and at each step extending all of them in all possible ways and keeping the n best results.

Discovering rules in this way was the brainchild of Ryszard Michalski, a Polish computer scientist. Michalski's hometown of Kalusz was successively part of Poland, Russia, Germany, and Ukraine, which may have left him more attuned than most to disjunctive concepts. After immigrating to the United States in 1970, he went on to found the symbolist school of machine learning, along with Tom Mitchell and Jaime Carbonell. He had an imperious personality. If you gave a talk at a machine-learning conference, the odds were good that at the end he'd raise his hand to point out that you had just rediscovered one of his old ideas.

Sets of rules are popular with retailers who are deciding which goods to stock. Typically, they use a more exhaustive approach than "divide and conquer," looking for all rules that strongly predict the purchase of each item. Walmart was a pioneer in this area. One of their early findings was that if you buy diapers you are also likely to buy beer. Huh? One interpretation of this is that Mom sends Dad to the supermarket to buy diapers, and as emotional compensation, Dad buys a case of beer to go with them. Knowing this, the supermarket can now sell more beer by putting it next to the diapers, which would never have occurred to it without rule mining. The "beer and diapers" rule has acquired legendary status among data miners (although some claim the legend is of the urban variety). Either way, it's a long way from the digital circuit design problems Michalski had in mind when he first started thinking about rule induction in the 1960s. When you invent a new learning algorithm, you can't even begin to imagine all the things it will be used for.

My first direct experience of rule learning in action was when, having just moved to the United States to start graduate school, I applied for a credit card. The bank sent me a letter saying "We regret that your application has been rejected due to INSUFFICIENT-TIME-AT-CURRENT-ADDRESS and NO-PREVIOUS-CREDIT-HISTORY" (or some other all-cap words to that effect). I knew right then that there was much research left to do in machine learning.

Between blindness and hallucination

Sets of rules are vastly more powerful than conjunctive concepts. They're so powerful, in fact, that you can represent *any* concept using them. It's not hard to see why. If you give me a complete list of all the instances of a concept, I can just turn each instance into a rule that specifies all attributes of that instance, and the set of all those rules is the definition of the concept. Going back to the dating example, one rule would be: *If it's a warm weekend night, there's nothing good on TV, and you propose going to a club, she'll say yes.* The table only contains a few examples, but if it contained all $2 \times 2 \times 2 \times 2 = 16$ possible ones, with each labeled "Date" or "No date," turning each positive example into a rule in this way would do the trick.

The power of rule sets is a double-edged sword. On the upside, you know you can always find a rule set that perfectly matches the data. But before you start feeling lucky, realize that you're at severe risk of finding a completely meaningless one. Remember the "no free lunch" theorem:

you can't learn without knowledge. And assuming that the concept can be defined by a set of rules is tantamount to assuming nothing.

An example of a useless rule set is one that just covers the exact positive examples you've seen and nothing else. This rule set looks like it's 100 percent accurate, but that's an illusion: it will predict that every new example is negative, and therefore get every positive one wrong. If there are more positive than negative examples overall, this will be even worse than flipping coins. Imagine a spam filter that decides an e-mail is spam only if it's an exact copy of a previously labeled spam message. It's easy to learn and looks great on the labeled data, but you might as well have no spam filter at all. Unfortunately, our "divide and conquer" algorithm could easily learn a rule set like that.

In his story "Funes the Memorious," Jorge Luis Borges tells of meeting a youth with perfect memory. This might at first seem like a great fortune, but it is in fact an awful curse. Funes can remember the exact shape of the clouds in the sky at an arbitrary time in the past, but he has trouble understanding that a dog seen from the side at 3:14 p.m. is the same dog seen from the front at 3:15 p.m. His own face in the mirror surprises him every time he sees it. Funes can't generalize; to him, two things are the same only if they look the same down to every last detail. An unrestricted rule learner is like Funes and is equally unable to function. Learning is forgetting the details as much as it is remembering the important parts. Computers are the ultimate idiot savants: they can remember everything with no trouble at all, but that's not what we want them to do.

The problem is not limited to memorizing instances wholesale. Whenever a learner finds a pattern in the data that is not actually true in the real world, we say that it has overfit the data. Overfitting is the central problem in machine learning. More papers have been written about it than about any other topic. Every powerful learner, whether symbolist, connectionist, or any other, has to worry about hallucinating patterns. The only safe way to avoid it is to severely restrict what the learner can learn, for example by requiring that it be a short conjunctive concept. Unfortunately, that throws out the baby with the bathwater, leaving the learner unable to see most of the true patterns that are visible in the data. Thus a good learner is forever walking the narrow path between blindness and hallucination.

Humans are not immune to overfitting, either. You could even say that it's the root cause of a lot of our evils. Consider the little white girl who, upon seeing a Latina baby at the mall, blurted out "Look, Mom, a baby maid!" (True event.) It's not that she's a natural-born bigot. Rather, she overgeneralized from the few Latina maids she has seen in her short life. The world is full of Latinas with other occupations, but she hasn't met them yet. Our beliefs are based on our experience, which gives us a very incomplete picture of the world, and it's easy to jump to false conclusions. Being smart and knowledgeable doesn't immunize you against overfitting, either. Aristotle overfit when he said that it takes a force to keep an object moving. Galileo's genius was to intuit that undisturbed objects keep moving without having visited outer space to witness it firsthand.

Learning algorithms are particularly prone to overfitting, though, because they have an almost unlimited capacity to find patterns in data. In the time it takes a human to find one pattern, a computer can find millions. In machine learning, the computer's greatest strength—its ability to process vast amounts of data and endlessly repeat the same steps without tiring—is also its Achilles' heel. And it's amazing what you can find if you search enough. *The Bible Code*, a 1998 bestseller, claimed that the Bible contains predictions of future events that you can find by skipping letters at regular intervals and assembling words from the letters you land on. Unfortunately, there are so many ways to do this that you're guaranteed to find "predictions" in any sufficiently long text. Skeptics replied by finding them in *Moby Dick* and Supreme Court rulings, along with mentions of Roswell and UFOs in Genesis. John von Neumann, one of the founding fathers of computer science, famously said that "with four parameters I can fit an elephant, and with five I can make him wiggle his trunk." Today we routinely learn models with millions of parameters, enough to give each elephant in the world his own distinctive wiggle. It's even been said that *data mining* means "torturing the data until it confesses."

Overfitting is seriously exacerbated by noise. Noise in machine learning just means errors in the data, or random events that you can't predict. Suppose that your friend really does like to go clubbing when there's nothing interesting on TV, but you misremembered occasion number 3 and wrote down that there *was* something good on TV that night. If you now try to come up with a set of rules that makes an exception for that night, you'll probably wind up with a worse answer than if you'd just ignored it. Or suppose that your friend had a hangover from going out the previous night and said no when ordinarily she would have said yes. Unless you know about the hangover, learning a set of rules that gets this example right is actually counterproductive: you're better off "misclassifying" it as a no. It gets worse: noise can make it impossible to come up with *any* consistent set of rules. Notice that occasions 2 and 3 are in fact indistinguishable: they have exactly the same attributes. If your friend said yes on occasion 2 and no on occasion 3, there's no rule that will get them both right.

Overfitting happens when you have too many hypotheses and not enough data to tell them apart. The bad news is that even for the simple conjunctive learner, the number of hypotheses grows exponentially with the number of attributes. Exponential growth is a scary thing. An *E. coli* bacterium can divide into two roughly every fifteen minutes; given enough nutrients it can grow into a mass of bacteria the size of Earth in about a day. When the number of things an algorithm needs to do grows exponentially with the size of its input, computer scientists call it a combinatorial explosion and run for cover. In machine learning, the number of possible instances of a concept is an exponential function of the number of attributes: if the attributes are Boolean, each new attribute doubles the number of possible instances by taking each previous instance and extending it with a yes or no for that attribute. In turn, the number of possible concepts is an exponential function of the number of possible instances: since a concept labels each instance as positive or negative, adding an instance doubles the number of possible concepts. As a result, the number of concepts is an exponential function of an exponential function of the number of attributes! In other words, machine learning is a combinatorial explosion of combinatorial explosions. Perhaps we should just give up and not waste our time on such a hopeless problem?

Fortunately, something happens in learning that kills off one of the exponentials, leaving only an "ordinary" singly exponential intractable problem. Suppose you have a bag full of concept definitions, each written on a piece of paper, and you take out a random one and see how well it matches the data. A bad definition is no more likely to get, say, all thousand examples in your data right than a coin is likely to come up heads a thousand times in a row. "A chair has four legs and is red or has a seat but no legs" will probably match some but not all chairs you've seen and also match some but not all other things. So if a random definition correctly matches a thousand examples, then it's extremely unlikely to be the wrong definition, or at least it's pretty close to the real one. And if the definition agrees with a million examples, then it's practically certain to be the right one. How else would it get all those examples right?

Of course, a real learning algorithm doesn't just take a random definition from the bag: it tries a vast number of them, and they're not chosen at random. The more definitions it tries, the more likely one of them will match all the examples just by chance. If you do more and more runs of a thousand coin flips, eventually it becomes practically certain that at least one run will come up all heads. And a learning algorithm doesn't need to explicitly try all definitions one by one; when it finds the best one, the result is the same as if it had tried all others. Since the number of definitions the algorithm considers can be as large as a doubly exponential function of the number of attributes, at some point it's practically guaranteed that the algorithm will find a bad hypothesis that looks good.

Bottom line: learning is a race between the amount of data you have and the number of hypotheses you consider. More data exponentially reduces the number of hypotheses that survive, but if you start with a lot of them, you may still have some bad ones left at the end. As a rule of thumb, if the learner only considers an exponential number of hypotheses (for example, all possible conjunctive concepts), then the data's exponential payoff cancels it and you're OK, provided you have plenty of examples and not too many attributes. On the other hand, if it considers a doubly exponential number (for example, all possible rule sets), then the data cancels only one of the exponentials and you're still in trouble. You can even figure out in advance how many examples you'll need to be pretty sure that the learner's chosen hypothesis is very close to the true one, provided it fits all the data; in other words, for the hypothesis to be probably approximately correct. Harvard's Leslie Valiant received the Turing Award, the Nobel Prize of computer science, for inventing this type of analysis, which he describes in his book entitled, appropriately enough, *Probably Approximately Correct*.

Accuracy you can believe in

In practice, Valiant-style analysis tends to be very pessimistic and to call for more data than you have. So how do you decide whether to believe what a learner tells you? Simple: you don't believe anything until you've verified it on data that *the learner didn't see*. If the patterns the learner hypothesized also hold true on new data, you can be pretty confident that they're real. Otherwise you know the learner overfit. This is just the scientific method applied to machine learning: it's not enough for a new theory to explain past evidence because it's easy to concoct a theory that does that; the theory must also make new predictions, and you only accept it after they've been experimentally verified. (And even then only provisionally, because future evidence could still falsify it.)

Einstein's general relativity was only widely accepted once Arthur Eddington empirically confirmed its prediction that the sun bends the light of distant stars. But you don't need to wait around for new data to arrive to decide whether you can trust your learner. Rather, you take the data you have and randomly divide it into a training set, which you give to the learner, and a test set, which you hide from it and use to verify its accuracy. Accuracy on held-out data is the gold standard in machine learning. You can write a paper about a great new learning algorithm you've invented, but if your algorithm is not significantly more accurate than previous ones on held-out data, the paper is not publishable.

Accuracy on previously unseen data is a pretty stringent test; so much so, in fact, that a lot of science fails it. That does not make it useless, because science is not just about prediction; it's also about explanation and understanding. But ultimately, if your models don't make accurate predictions on new data, you can't be sure you've truly understood or explained the underlying phenomena. And for machine learning, testing on unseen data is indispensable because it's the only way to tell whether the learner has overfit or not.

Even test-set accuracy is not foolproof. According to legend, in an early military application a simple learner detected tanks with 100 percent accuracy in both the training set and the test set, each consisting of one hundred images. Amazing—or suspicious? Turns out all the tank images were lighter than the nontank ones, and that's all the learner was picking up. These days we have larger data sets, but the quality of data collection isn't necessarily better, so caveat emptor. Hard-nosed empirical evaluation played an important role in the growth of machine learning from a fledgling field into a mature one. Up to the late 1980s, researchers in each tribe mostly believed their own rhetoric, assumed their paradigm was fundamentally better, and communicated little with the other camps. Then symbolists like Ray Mooney and Jude Shavlik started to systematically compare the different algorithms on the same data sets and—surprise, surprise—no clear winner emerged. Today the rivalry continues, but there is much more cross-pollination. Having a common experimental framework and a large repository of data sets maintained by the machine-learning group at the University of California, Irvine, did wonders for progress. And as we'll see, our best hope of creating a universal learner lies in synthesizing ideas from different paradigms.

Of course, it's not enough to be able to tell when you're overfitting; we need to avoid it in the first place. That means stopping short of perfectly fitting the data even if we're able to. One method is to use statistical significance tests to make sure the patterns we're seeing are really there. For example, a rule covering three hundred positive examples versus one hundred negatives and a rule covering three positives versus one negative are both 75 percent accurate on the training data, but the first rule is almost certainly better than coin flipping, while the second isn't, since four flips of an unbiased coin could easily result in three heads. When constructing a rule, if at some point we can't find any conditions that significantly improve its accuracy then we just stop, even if it still covers some negative examples. This reduces the rule's training-set accuracy, but probably makes it a more accurate generalization, which is what we really care about.

We're not home free yet, though. If I try one rule and it's 75 percent accurate on four hundred examples, I can probably believe it. But if I try a million rules and the best one is 75 percent accurate on four hundred examples, I probably can't, because that could easily happen by chance. This is the same problem you have when picking a mutual fund. The Clairvoyant Fund just beat the market ten years in a row. Wow, the manager must be a genius. Or not? If you have a thousand funds to choose from, the odds are better than even that one will beat the market ten years in a row, even if they're all secretly run by dart-throwing monkeys. The scientific literature is also plagued by this problem. Significance tests are the gold standard for deciding whether a research result is publishable, but if several teams look for an effect and only one finds it, chances are it didn't, even though you'd never guess that from reading their solid-looking paper. One solution would be to also publish negative results, so you'd know about all those failed attempts, but that hasn't caught on. In machine learning, we can keep track of how many rules we've tried and adjust our significance tests accordingly, but then they tend to throw out a lot of good rules along with the bad ones. A better method is to realize that some false hypotheses will inevitably get through, but keep their number under control by rejecting enough low-significance ones, and then test the surviving hypotheses on further data.

Another popular method is to prefer simpler hypotheses. The "divide and conquer" algorithm implicitly prefers simpler rules because it stops adding conditions to a rule as soon as it covers only positive examples and stops adding rules as soon as all positive examples are covered. But to combat overfitting, we need a stronger preference for simpler rules, one that will cause us to stop adding conditions even before all negative examples have been covered. For example, we can subtract a penalty proportional to the length of the rule from its accuracy and use that as an evaluation measure.

The preference for simpler hypotheses is popularly known as Occam's razor, but in a machine-learning context this is somewhat misleading. "Entities should not be multiplied beyond necessity," as the razor is often paraphrased, just means choosing the simplest theory that fits the data. Occam would probably have been perplexed by the notion that we should prefer a theory that does *not* perfectly account for the evidence on the grounds that it will generalize better. Simple theories are preferable because they incur a lower cognitive cost (for us) and a lower computational cost (for our algorithms), not because we necessarily expect them to be more accurate. On the contrary, even our most elaborate models are usually oversimplifications of reality. Even among theories that perfectly fit the data, we know from the "no free lunch" theorem that there's no guarantee that the simplest one will generalize best, and in fact some of the best learning algorithms—like boosting and support vector machines—learn what appear to be gratuitously complex models. (We'll see why they work in Chapters 7 and 9.)

If your learner's test-set accuracy disappoints, you need to diagnose the problem. Was it blindness or hallucination? In machine learning, the technical terms for these are *bias* and *variance*. A clock that's always an hour late has high bias but low variance. If instead the clock alternates erratically between fast and slow but on average tells the right time, it has high variance but low bias. Suppose you're down at the pub with some friends, drinking and playing darts. Unbeknownst to them, you've been practicing for years, and you're a master of the game. All your darts go straight to the bull's-eye. You have low bias and low variance, which is shown in the bottom left corner of this diagram:

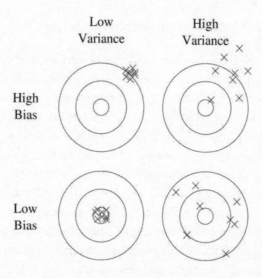

Your friend Ben is also pretty good, but he's had a bit too much to drink. His darts are all over, but he loudly points out that on average he's hitting the bull's-eye. (Maybe he should have been a statistician.) This is the low-bias, high-variance case, shown in the bottom right corner. Ben's girlfriend, Ashley, is very steady, but she has a tendency to aim too high and to the right. She has low variance and high bias (top left corner). Cody, who's visiting from out of town and has never played darts before, is both all over and off center. He has both high bias and high variance (top right).

You can estimate the bias and variance of a learner by comparing its predictions after learning on random variations of the training set. If it keeps making the same mistakes, the problem is bias, and you need a more flexible learner (or just a different one). If there's no pattern to the mistakes, the problem is variance, and you want to either try a less flexible learner or get more data. Most learners have a knob you can turn to make them more or less flexible, such as the threshold for significance tests or the penalty on the size of the model. Tweaking that knob is your first resort.

Induction is the inverse of deduction

The deeper problem, however, is that most learners start out knowing too little, and no amount of knob-twiddling will get them to the finish line. Without the guidance of an adult brain's worth of knowledge, they can easily go astray. Even though it's what most learners do, just assuming you know the form of the truth (for example, that it's a small set of rules) is not much to hang your hat on. A strict empiricist would say that that's all a newborn has, encoded in her brain's architecture, and indeed children overfit more than adults do, but we would like to learn faster than a child does. (Eighteen years is a long time, and that's not counting college.) The Master Algorithm should be able to start with a large body of knowledge, whether it was provided by humans or learned in previous runs, and use it to guide new generalizations from data. That's what scientists do, and it's as far as it gets from a blank slate. The "divide and conquer" rule induction algorithm can't do it, but there's another way to learn rules that can.

The key is to realize that induction is just the inverse of deduction, in the same way that subtraction is the inverse of addition, or integration the inverse of differentiation. This idea was first proposed by William Stanley Jevons in the late 1800s. Steve Muggleton and Wray Buntine, an English Australian team, designed the first practical algorithm based on it in 1988. The strategy of taking a well-known operation and figuring out its inverse has a storied history in mathematics. Applying it to addition led to the invention of the integers, because without negative numbers, addition doesn't always have an inverse $(3 - 4 = -1)$. Similarly, applying it to multiplication led to the rationals, and applying it to squaring led to complex numbers. Let's see if we can apply it to deduction. A classic example of deductive reasoning is:

Socrates is human.
All humans are mortal.
Therefore. . . . . .? . . .

The first statement is a fact about Socrates, and the second is a general rule about humans. What follows? That Socrates is mortal, of course, by applying the rule to Socrates. In inductive reasoning we start instead with the initial and derived facts, and look for a rule that would allow us to infer the latter from the former:

Socrates is human.
......?...
Therefore Socrates is mortal.

One such rule is: *If Socrates is human, then he's mortal.* This does the job, but is not very useful because it's specific to Socrates. But now we apply Newton's principle and generalize the rule to all entities: *If an entity is human, then it's mortal.* Or, more succinctly: *All humans are mortal.* Of course, it would be rash to induce this rule from Socrates alone, but we know similar facts about other humans:

Plato is human. Plato is mortal.
Aristotle is human. Aristotle is mortal.
And so on.

For each pair of facts, we construct the rule that allows us to infer the second fact from the first one and generalize it by Newton's principle. When the same general rule is induced over and over again, we can have some confidence that it's true.

So far we haven't done anything that the "divide and conquer" algorithm couldn't do. Suppose, however, that instead of knowing that Socrates, Plato, and Aristotle are human, we just know that they're philosophers. We still want to conclude that they're mortal, and we have previously induced or been told that all humans are mortal. What's missing now? A different rule: *All philosophers are human.* This also a valid generalization (at least until we solve AI and robots start philosophizing), and it "fills the hole" in our reasoning:

Socrates is a philosopher.
All philosophers are human.
All humans are mortal.
Therefore Socrates is mortal.

We can also induce rules purely from other rules. If we know that all philosophers are human and mortal, we can induce that all humans are mortal. (We don't induce that all mortals are human because we know other mortal creatures, like cats and dogs. On the other hand, scientists, artists, and so on are also human and mortal, reinforcing the rule.) In general, the more rules and facts we start out with, the more opportunities we have to induce new rules using "inverse deduction." And the more rules we induce, the more rules we can induce. It's a virtuous circle of knowledge creation, limited only by overfitting risk and computational cost. But here, too, having initial knowledge helps: if instead of one large hole we have many small ones to fill, our induction steps will be less risky and therefore less likely to overfit. (For example, given the same number of examples, inducing that all philosophers are human is less risky than inducing that all humans are mortal.)

Inverting an operation is often difficult because the inverse is not unique. For example, a positive number has two square roots, one positive and one negative ($2^2 = (-2)^2 = 4$). Most famously, integrating the derivative of a function only recovers the function up to a constant. The derivative of a function tells us how much that function goes up or down at each point. Adding up all those changes gives us the function back, except we don't know where it started; we can "slide" the integrated function up or down without changing the derivative. To make life easy, we can "clamp down" the function by assuming the additive constant is zero. Inverse deduction has a similar problem, and Newton's principle is one solution. For example, from *All Greek philosophers are human* and *All Greek philosophers are mortal* we can induce that *All humans are mortal*, or just that *All Greeks are mortal*. But why settle for the more modest generalization? Instead, we can assume that all humans are mortal until we meet an exception. (Which, according to Ray Kurzweil, will be soon.)

In the meantime, one important application of inverse deduction is predicting whether new drugs will have harmful side effects. Failure during animal testing and clinical trials is the main reason new drugs take many years and billions of dollars to develop. By generalizing from known toxic molecular structures, we can form rules that quickly weed out many apparently promising compounds, greatly increasing the chances of successful trials on the remaining ones.

Learning to cure cancer

More generally, inverse deduction is a great way to discover new knowledge in biology, and doing that is the first step in curing cancer. According to the Central Dogma, everything that happens in a living cell is ultimately controlled by its genes, via the proteins whose synthesis they initiate. In effect, a cell is like a tiny computer, and DNA is the program running on it: change the DNA, and a skin cell can become a neuron or a mouse cell can turn into a human one. In a computer program, all bugs are the programmer's fault. But in a cell, bugs can arise spontaneously, when radiation or a copying error changes a gene into a different one, a gene is accidentally copied twice, and so on. Most of the time these mutations cause the cell to die silently, but sometimes the cell starts to grow and divide uncontrollably and a cancer is born.

Curing cancer means stopping the bad cells from reproducing without harming the good ones. That requires knowing how they differ, and in particular how their genomes differ, since all else follows from that. Luckily, gene sequencing is becoming routine and affordable. Using it, we can learn to predict which drugs will work against which cancer genes. This contrasts with traditional chemotherapy, which affects all cells indiscriminately. Learning which drugs work against which mutations requires a database of patients, their cancers' genomes, the drugs tried, and the outcomes. The simplest rules encode one-to-one correspondences between genes and drugs, such as *If the BCR-ABL gene is present, then use Gleevec.* (BCR-ABL causes a type of leukemia, and Gleevec cures it in nine out of ten patients.) Once sequencing cancer genomes and collating treatment outcomes becomes standard practice, many more rules like this will be discovered.

That's only the beginning, however. Most cancers involve a combination of mutations, or can only be cured by drugs that haven't been discovered yet. The next step is to learn rules with more complex conditions, involving the cancer's genome, the patient's genome and medical history, known side effects of drugs, and so on. But ultimately what we need is a model of how the entire cell works, enabling us to simulate on the computer the effect of a specific patient's mutations, as well as the effect of different combinations of drugs, existing or speculative. Our main sources of information for building such models are DNA sequencers, gene expression microarrays, and the biological literature. Combining these is where inverse deduction can shine.

Adam, the robot scientist we met in Chapter 1, gives a preview. Adam's goal is to figure out how yeast cells work. It starts with basic knowledge of yeast genetics and metabolism and a trove of gene expression data from yeast cells. It then uses inverse deduction to hypothesize which genes are expressed as which proteins, designs microarray experiments to test them, revises its hypotheses, and repeats. Whether each gene is expressed depends on other genes and conditions in the environment, and the resulting web of interactions can be represented as a set of rules, such as:

If the temperature is high, gene A is expressed.
If gene A is expressed and gene B is not, gene C is expressed.
If gene C is expressed, gene D is not.

If we knew the first and third rules but not the second, and we had microarray data where at a high temperature B and D were not expressed, we could induce the second rule by inverse deduction. Once we have that rule, and perhaps have verified it using a microarray experiment, we can use it as the basis for further inductive inferences. In a similar manner, we can piece together the sequences of chemical reactions by which proteins do their work.

Just knowing which genes regulate which genes and how proteins organize the cell's web of chemical reactions is not enough, though. We also need to know how much of each molecular species is produced. DNA microarrays and other experiments can provide this type of quantitative information, but inverse deduction, with its "all or none" logical character, is not very good at dealing with it. For that we need the connectionist methods that we'll meet in the next chapter.

A game of twenty questions

Another limitation of inverse deduction is that it's very computationally intensive, which makes it hard to scale to massive data sets. For these, the symbolist algorithm of choice is decision tree induction. Decision trees can be viewed as an answer to the question of what to do if rules of more than one concept match an instance. How do we then decide which concept the instance belongs to? If we see a partly occluded object with a flat surface and four legs, how do we decide whether it is a table or a chair? One option is to order the rules, for example by decreasing accuracy, and choose the first one that matches. Another is to let the rules vote. Decision trees instead ensure a priori that each instance will be matched by exactly one rule. This will be the case if each pair of rules differs in at least one attribute test, and such a rule set can be organized into a decision tree. For example, consider these rules:

> *If you're for cutting taxes and pro-life, you're a Republican.*
> *If you're against cutting taxes, you're a Democrat.*
> *If you're for cutting taxes, pro-choice, and against gun control, you're an*
> *   independent.*
> *If you're for cutting taxes, pro-choice, and pro-gun control, you're a*
> *   Democrat.*

These can be organized into the following decision tree:

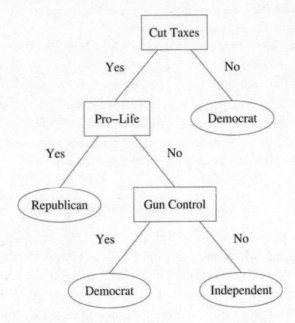

A decision tree is like playing a game of twenty questions with an instance. Starting at the root, each node asks about the value of one attribute, and depending on the answer, we follow one or another branch. When we arrive at a leaf, we read off the predicted concept. Each path from the root to a leaf corresponds to a rule. If this reminds you of those annoying phone menus you have to get through when you call customer service, it's not an accident: a phone menu is a decision tree. The computer on the other end of the line is playing a game of twenty questions with you to figure out what you want, and each menu is a question.

According to the decision tree above, you're either a Republican, a Democrat, or an independent; you can't be more than one, or none of the above. Sets of concepts with this property are called sets of classes, and the algorithm that predicts them is a classifier. A single concept implicitly defines two classes: the concept itself and its negation. (For example, spam and nonspam.) Classifiers are the most widespread form of machine learning.

We can learn decision trees using a variant of the "divide and conquer" algorithm. First we pick an attribute to test at the root. Then we focus on the examples that went down each branch and pick the next test for those. (For example, we check whether tax-cutters are pro-life or pro-choice.) We repeat this for each new node we induce until all the examples in a branch have the same class, at which point we label that branch with the class.

One salient question is how to pick the best attribute to test at a node. Accuracy—the number of correctly predicted examples—doesn't work very well, because we're not trying to predict a particular class; rather, we're trying to gradually separate the classes until each branch is "pure." This brings to mind the concept of entropy from information theory. The entropy of a set of objects is a measure of the amount of disorder in it. If a group of 150 people includes 50 Republicans, 50 Democrats, and 50 independents, its political entropy is maximum. On the other hand, if they're all Republican then the entropy is zero (as far as party affiliation goes). So to learn a good decision tree, we pick at each node the attribute that on average yields the lowest class entropy across all its branches, weighted by how many examples go into each branch.

As with rule learning, we don't want to induce a tree that perfectly predicts the classes of all the training examples, because it would probably overfit. As before, we can use significance tests or a penalty on the size of the tree to prevent this.

Having a branch for each value of an attribute is fine if the attribute is discrete, but what about numeric attributes? If we had a branch for every value of a continuous variable, the tree would be infinitely wide. A simple solution is to pick a few key thresholds by entropy and use those. For example, is the patient's temperature above or below 100 degrees Fahrenheit? That, combined with other symptoms, may be all the doctor needs to know about the patient's temperature to decide if he has an infection.

Decision trees are used in many different fields. In machine learning, they grew out of work in psychology. Earl Hunt and colleagues used them in the 1960s to model how humans acquire new concepts, and one of Hunt's graduate students, J. Ross Quinlan, later tried using them for chess. His original goal was to predict the outcome of king-rook versus king-knight endgames from the board positions. From those humble beginnings, decision trees have grown to be, according to surveys, the most widely used machine-learning algorithm. It's not hard to see why: they're easy to understand, fast to learn, and usually quite accurate without too much tweaking. Quinlan is the most prominent researcher in the symbolist school. An unflappable, down-to-earth Australian, he made decision trees the gold standard in classification by dint of relentlessly improving them year after year, and writing beautifully clear papers about them.

Whatever you want to predict, there's a good chance someone has used a decision tree for it. Microsoft's Kinect uses decision trees to figure out where various parts of your body are from the output of its depth camera; it can then use their motions to control the Xbox game console. In a 2002 head-to-head competition, decision trees correctly predicted three out of every four Supreme Court rulings, while a panel of experts got less than 60 percent correct. Thousands of decision tree users can't be wrong, you think, and sketch one to predict your friend's reply when you ask her out:

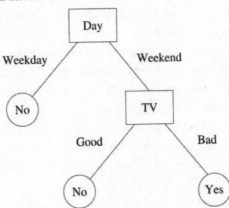

According to this tree, tonight she'll say yes. With a deep breath, you pick up the phone and dial her number.

The symbolists

The symbolists' core belief is that all intelligence can be reduced to manipulating symbols. A mathematician solves equations by moving symbols around and replacing symbols by other symbols according to predefined rules. The same is true of a logician carrying out deductions. According to this hypothesis, intelligence is independent of the substrate; it doesn't matter if the symbol manipulations are done by writing on a blackboard, switching transistors on and off, firing neurons, or playing with Tinkertoys. If you have a setup with the power of a universal Turing machine, you can do anything. Software can be cleanly separated from hardware, and if your concern is figuring out how machines can learn, you (thankfully) don't need to worry about the latter beyond buying a PC or cycles on Amazon's cloud.

Symbolist machine learners share this belief in the power of symbol manipulation with many other computer scientists, psychologists, and philosophers. The psychologist David Marr argued that every information processing system should be studied at three distinct levels: the fundamental properties of the problem it's solving; the algorithms and representations used to solve it; and how they are physically implemented. For example, addition can be defined by a set of axioms irrespective of how it's carried out; numbers can be expressed in different ways (e.g., Roman and Arabic) and added using different algorithms; and these can be implemented using an abacus, a pocket calculator, or even, very inefficiently, in your head. Learning is a prime example of a cognitive faculty we can profitably study according to Marr's levels.

Symbolist machine learning is an offshoot of the knowledge engineering school of AI. In the 1970s, so-called knowledge-based systems scored some impressive successes, and in the 1980s they spread rapidly, but then they died out. The main reason they did was the infamous knowledge acquisition bottleneck: extracting knowledge from experts and encoding it as rules is just too difficult, labor-intensive, and fail-ure-prone to be viable for most problems. Letting the computer auto-matically learn to, say, diagnose diseases by looking at databases of past patients' symptoms and the corresponding outcomes turned out to be much easier than endlessly interviewing doctors. Suddenly, the work of pioneers like Ryszard Michalski, Tom Mitchell, and Ross Quinlan had a new relevance, and the field hasn't stopped growing since. (Another im-portant problem was that knowledge-based systems had trouble dealing with uncertainty, of which more in Chapter 6.)

Because of its origins and guiding principles, symbolist machine learning is still closer to the rest of AI than the other schools. If com-puter science were a continent, symbolist learning would share a long border with knowledge engineering. Knowledge is traded in both direc-tions—manually entered knowledge for use in learners, induced knowl-edge for addition to knowledge bases—but at the end of the day the rationalist-empiricist fault line runs right down that border, and cross-ing it is not easy.

Symbolism is the shortest path to the Master Algorithm. It doesn't require us to figure out how evolution or the brain works, and it avoids the mathematical complexities of Bayesianism. Sets of rules and deci-sion trees are easy to understand, so we know what the learner is up to. This makes it easier to figure out what it's doing right and wrong, fix the latter, and have confidence in the results.

Despite the popularity of decision trees, inverse deduction is the better starting point for the Master Algorithm. It has the crucial prop-erty that incorporating knowledge into it is easy—and we know Hume's problem makes that essential. Also, sets of rules are an exponentially more compact way to represent most concepts than decision trees. Converting a decision tree to a set of rules is easy: each path from the root to a leaf becomes a rule, and there's no blowup. On the other hand, in the worst case converting a set of rules into a decision tree requires converting each rule into a mini-decision tree, and then replacing each leaf of rule 1's tree with a copy of rule 2's tree, each leaf of each copy of rule 2 with a copy of rule 3, and so on, causing a massive blowup.

Inverse deduction is like having a superscientist systematically looking at the evidence, considering possible inductions, collating the strongest, and using those along with other evidence to construct yet further hypotheses—all at the speed of computers. It's clean and beautiful, at least for the symbolist taste. On the other hand, it has some serious shortcomings. The number of possible inductions is vast, and unless we stay close to our initial knowledge, it's easy to get lost in space. Inverse deduction is easily confused by noise: how do we figure out what the missing deductive steps are, if the premises or conclusions are themselves wrong? Most seriously, real concepts can seldom be concisely defined by a set of rules. They're not black and white: there's a large gray area between, say, spam and nonspam. They require weighing and accumulating weak evidence until a clear picture emerges. Diagnosing an illness involves giving more weight to some symptoms than others, and being OK with incomplete evidence. No one has ever succeeded in learning a set of rules that will recognize a cat by looking at the pixels in an image, and probably no one ever will.

Connectionists, in particular, are highly critical of symbolist learning. According to them, concepts you can define with logical rules are only the tip of the iceberg; there's a lot going on under the surface that formal reasoning just can't see, in the same way that most of what goes on in our minds is subconscious. You can't just build a disembodied automated scientist and hope he'll do something meaningful—you have to first endow him with something like a real brain, connected to real senses, growing up in the world, perhaps even stubbing his toe every now and then. And how do you build such a brain? By reverse engineering the competition. If you want to reverse engineer a car, you look under the hood. If you want to reverse engineer the brain, you look inside the skull.

CHAPTER FOUR

How Does Your Brain Learn?

Hebb's rule, as it has come to be known, is the cornerstone of connectionism. Indeed, the field derives its name from the belief that knowledge is stored in the connections between neurons. Donald Hebb, a Canadian psychologist, stated it this way in his 1949 book *The Organization of Behavior*: "When an axon of cell *A* is near enough cell *B* and repeatedly or persistently takes part in firing it, some growth process or metabolic change takes place in one or both cells such that *A*'s efficiency, as one of the cells firing *B*, is increased." It's often paraphrased as "Neurons that fire together wire together."

Hebb's rule was a confluence of ideas from psychology and neuroscience, with a healthy dose of speculation thrown in. Learning by association was a favorite theme of the British empiricists, from Locke and Hume to John Stuart Mill. In his *Principles of Psychology*, William James enunciates a general principle of association that's remarkably similar to Hebb's rule, with neurons replaced by brain processes and firing efficiency by propagation of excitement. Around the same time, the great Spanish neuroscientist Santiago Ramón y Cajal was making the first detailed observations of the brain, staining individual neurons using the recently invented Golgi method and cataloguing what he saw like a botanist classifying new species of trees. By Hebb's time, neuroscientists had a rough understanding of how neurons work, but he was the first to propose a mechanism by which they could encode associations.

In symbolist learning, there is a one-to-one correspondence between symbols and the concepts they represent. In contrast, connectionist representations are distributed: each concept is represented by many neurons, and each neuron participates in representing many different concepts. Neurons that excite one another form what Hebb called a cell assembly. Concepts and memories are represented in the brain by cell assemblies. Each of these can include neurons from different brain regions and overlap with other assemblies. The cell assembly for "leg" includes the one for "foot," which includes assemblies for the image of a foot and the sound of the word *foot*. If you ask a symbolist system where the concept "New York" is represented, it can point to the precise location in memory where it's stored. In a connectionist system, the answer is "it's stored a little bit everywhere."

Another difference between symbolist and connectionist learning is that the former is sequential, while the latter is parallel. In inverse deduction, we figure out one step at a time what new rules are needed to arrive at the desired conclusion from the premises. In connectionist models, all neurons learn simultaneously according to Hebb's rule. This mirrors the different properties of computers and brains. Computers do everything one small step at a time, like adding two numbers or flipping a switch, and as a result they need a lot of steps to accomplish anything useful; but those steps can be very fast, because transistors can switch on and off billions of times per second. In contrast, brains can perform a large number of computations in parallel, with billions of neurons working at the same time; but each of those computations is slow, because neurons can fire at best a thousand times per second.

The number of transistors in a computer is catching up with the number of neurons in a human brain, but the brain wins hands down in the number of connections. In a microprocessor, a typical transistor is directly connected to only a few others, and the planar semiconductor technology used severely limits how much better a computer can do. In contrast, a neuron has thousands of synapses. If you're walking down the street and come across an acquaintance, it takes you only about a tenth of a second to recognize her. At neuron switching speeds, this is barely enough time for a hundred processing steps, but in those hundred steps your brain manages to scan your entire memory, find the best match, and adapt it to the new context (different clothes, different lighting, and so on). In a brain, each processing step can be very complex and involve a lot of information, consonant with a distributed representation.

This does not mean that we can't simulate a brain with a computer; after all, that's what connectionist algorithms do. Because a computer is a universal Turing machine, it can implement the brain's computations as well as any others, provided we give it enough time and memory. In particular, the computer can use speed to make up for lack of connectivity, using the same wire a thousand times over to simulate a thousand wires. In fact, these days the main limitation of computers compared to brains is energy consumption: your brain uses only about as much power as a small lightbulb, while Watson's supply could light up a whole office building.

To simulate a brain, we need more than Hebb's rule, however; we need to understand how the brain is built. Each neuron is like a tiny tree, with a prodigious number of roots—the dendrites—and a slender, sinuous trunk—the axon. The brain is a forest of billions of these trees, but there's something unusual about them. Each tree's branches make connections—synapses—to the roots of thousands of others, forming a massive tangle like nothing you've ever seen. Some neurons have short axons and some have exceedingly long ones, reaching clear from one side of the brain to the other. Placed end to end, the axons in your brain would stretch from Earth to the moon.

And this jungle crackles with electricity. Sparks run along tree trunks and set off more sparks in neighboring trees. Every now and then, a whole area of the jungle whips itself into a frenzy before settling down again. When you wiggle your toe, a series of electric discharges, called action potentials, runs all the way down your spinal chord and leg until it reaches your toe muscles and tells them to move. Your brain at work is a symphony of these electric sparks. If you could sit inside it and watch what happens as you read this page, the scene you'd see would make even the busiest science-fiction metropolis look laid back by comparison. The end result of this phenomenally complex pattern of neuron firings is your consciousness.

In Hebb's time there was no way to measure synaptic strength or change in it, let alone figure out the molecular biology of synaptic change. Today, we know that synapses do grow (or form anew) when the postsynaptic neuron fires soon after the presynaptic one. Like all cells, neurons have different concentrations of ions inside and outside, creating a voltage across their membrane. When the presynaptic neuron fires, tiny sacs release neurotransmitter molecules into the synaptic cleft. These cause channels in the postsynaptic neuron's membrane to open, letting in potassium and sodium ions and changing the voltage across the membrane as a result. If enough presynaptic neurons fire close together, the voltage suddenly spikes, and an action potential travels down the postsynaptic neuron's axon. This also causes the ion channels to become more responsive and new channels to appear, strengthening the synapse. To the best of our knowledge, this is how neurons learn.

The next step is to turn it into an algorithm.

The rise and fall of the perceptron

The first formal model of a neuron was proposed by Warren McCulloch and Walter Pitts in 1943. It looked a lot like the logic gates computers are made of. An OR gate switches on when at least one of its inputs is on, and an AND gate when all of them are on. A McCulloch-Pitts neuron switches on when the number of its active inputs passes some threshold. If the threshold is one, the neuron acts as an OR gate; if the threshold is equal to the number of inputs, as an AND gate. In addition, a McCulloch-Pitts neuron can prevent another from switching on, which models both inhibitory synapses and NOT gates. So a network of neurons can do all the operations a computer does. In the early days, computers were often called electronic brains, and this was not just an analogy.

What the McCulloch-Pitts neuron doesn't do is learn. For that we need to give variable weights to the connections between neurons, resulting in what's called a perceptron. Perceptrons were invented in the late 1950s by Frank Rosenblatt, a Cornell psychologist. A charismatic speaker and lively character, Rosenblatt did more than anyone else to shape the early days of machine learning. The name *perceptron* derives from his interest in applying his models to perceptual tasks like speech and character recognition. Rather than implement perceptrons in software, which was very slow in those days, Rosenblatt built his own devices. The weights were implemented by variable resistors like those found in dimmable light switches, and weight learning was carried out by electric motors that turned the knobs on the resistors. (Talk about high tech!)

In a perceptron, a positive weight represents an excitatory connection, and a negative weight an inhibitory one. The perceptron outputs 1 if the weighted sum of its inputs is above threshold, and 0 if it's below. By varying the weights and threshold, we can change the function that the perceptron computes. This ignores a lot of the details of how neurons work, of course, but we want to keep things as simple as possible; our goal is to develop a general-purpose learning algorithm, not to build a realistic model of the brain. If some of the details we ignored turn out to be important, we can always add them in later. Despite our simplifying abstractions, however, we can still see how each part of this model corresponds to a part of the neuron:

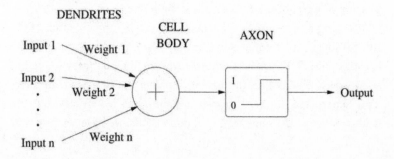

The higher an input's weight, the stronger the corresponding synapse. The cell body adds up all the weighted inputs, and the axon applies a step function to the result. The axon's box in the diagram shows the graph of a step function: 0 for low values of the input, abruptly changing to 1 when the input reaches the threshold.

Suppose a perceptron has two continuous inputs x and y. (In other words, x and y can take on any numeric values, not just 0 and 1.) Then each example can be represented by a point on the plane, and the boundary between positive examples (for which the perceptron outputs 1) and negative ones (output 0) is a straight line:

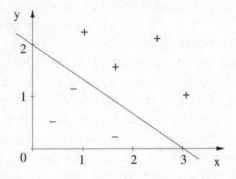

This is because the boundary is the set of points where the weighted sum exactly equals the threshold, and a weighted sum is a linear function. For example, if the weights are 2 for x and 3 for y and the threshold is 6, the boundary is defined by the equation $2x + 3y = 6$. The point $x = 0$, $y = 2$ is on the boundary, and to stay on it we have to take three steps across for every two steps down, so that the gain in x makes up for the loss in y. The resulting points form a straight line.

Learning a perceptron's weights means varying the direction of the straight line until all the positive examples are on one side and all the negative ones on the other. In one dimension, the boundary is a point; in two, it's a straight line; in three, it's a plane; and in more than three, it's a hyperplane. It's hard to visualize things in hyperspace, but the math works just the same way. In n dimensions, we have n inputs and the perceptron has n weights. To decide whether the perceptron fires or not, we multiply each weight by the corresponding input and compare the sum of all of them with the threshold.

If all inputs have a weight of one and the threshold is half the number of inputs, then the perceptron fires if more than half its inputs fire. In other words, the perceptron is a like a tiny parliament where the majority wins. (Or perhaps not so tiny, considering it can have thousands of members.) It's not altogether democratic, though, because in general not everyone has an equal vote. A neural network is more like a social network, where a few close friends count for more than thousands of Facebook ones. And it's the friends you trust most that influence you the most. If a friend recommends a movie and you go see it and like it, next time around you'll probably follow her advice again. On the other hand, if she keeps gushing about movies you didn't enjoy, you will start to ignore her opinions (and perhaps your friendship even wanes a bit).

This is how Rosenblatt's perceptron algorithm learns weights.

Consider the grandmother cell, a favorite thought experiment of cognitive neuroscientists. The grandmother cell is a neuron in your brain that fires whenever you see your grandmother, and only then. Whether or not grandmother cells really exist is an open question, but let's design one for use in machine learning. A perceptron learns to recognize your grandmother as follows. The inputs to the cell are either the raw pixels in the image or various hardwired features of it, like *brown eyes*, which takes the value 1 if the image contains a pair of brown eyes and 0 otherwise. In the beginning, all the connections from features to the neuron have small random weights, like the synapses in your brain at birth. Then we show the perceptron a series of images, some of your grandmother and some not. If it fires upon seeing an image of your grandmother, or doesn't fire upon seeing something else, then no learning needs to happen. (If it ain't broke, don't fix it.) But if the perceptron fails to fire when it's looking at your grandmother, that means the weighted sum of its inputs should have been higher, so we increase the weights of the inputs that are on. (For example, if your grandmother has brown eyes, the weight of that feature goes up.) Conversely, if the perceptron fires when it shouldn't, we decrease the weights of the active inputs. It's the errors that drive the learning. Over time, the features that are indicative of your grandmother acquire high weights, and the ones that aren't get low weights. Once the perceptron always fires upon seeing your grandmother, and only then, the learning is complete.

The perceptron generated a lot of excitement. It was simple, yet it could recognize printed letters and speech sounds just by being trained with examples. A colleague of Rosenblatt's at Cornell proved that, if the positive and negative examples could be separated by a hyperplane, the perceptron would find it. For Rosenblatt and others, a genuine understanding of how the brain learns seemed within reach, and with it a powerful general-purpose learning algorithm.

But then the perceptron hit a brick wall. The knowledge engineers were irritated by Rosenblatt's claims and envious of all the attention and funding neural networks, and perceptrons in particular, were getting. One of them was Marvin Minsky, a former classmate of Rosenblatt's at the Bronx High School of Science and by then the leader of the AI group at MIT. (Ironically, his PhD had been on neural networks, but he had grown disillusioned with them.) In 1969, Minsky and his colleague Seymour Papert published *Perceptrons*, a book detailing the shortcomings of the eponymous algorithm, with example after example of simple things it couldn't learn. The simplest one—and therefore the most damning—was the exclusive-OR function, or XOR for short, which is true if one of its inputs is true but not both. For example, Nike's two most loyal demographics are supposedly teenage boys and middle-aged women. In other words, you're likely to buy Nike shoes if you're young XOR female. Young is good, female is good, but both is not. You're also an unpromising target for Nike advertising if you're neither young nor female. The problem with XOR is that there is no straight line capable of separating the positive from the negative examples. This figure shows two failed candidates:

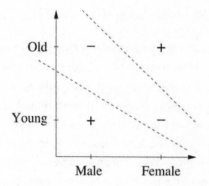

Since perceptrons can only learn linear boundaries, they can't learn XOR. And if they can't do even that, they're not a very good model of how the brain learns, or a viable candidate for the Master Algorithm.

A perceptron models only a single neuron's learning, however, and although Minsky and Papert acknowledged that layers of interconnected neurons should be capable of more, they didn't see a way to learn them. Neither did anyone else. The problem is that there's no clear way to change the weights of the neurons in the "hidden" layers to reduce the errors made by the ones in the output layer. Every hidden neuron influences the output via multiple paths, and every error has a thousand fathers. Who do you blame? Or, conversely, who gets the credit for correct outputs? This credit-assignment problem shows up whenever we try to learn a complex model and is one of the central problems in machine learning.

Perceptrons was mathematically unimpeachable, searing in its clarity, and disastrous in its effects. Machine learning at the time was associated mainly with neural networks, and most researchers (not to mention funders) concluded that the only way to build an intelligent system was to explicitly program it. For the next fifteen years, knowledge engineering would hold center stage, and machine learning seemed to have been consigned to the ash heap of history.

Physicist makes brain out of glass

If the history of machine learning were a Hollywood movie, the villain would be Marvin Minsky. He's the evil queen who gives Snow White a poisoned apple, leaving her in suspended animation. (In a 1988 essay, Seymour Papert even compared himself, tongue-in-cheek, to the huntsman the queen sent to kill Snow White in the forest.) And Prince Charming would be a Caltech physicist by the name of John Hopfield. In 1982, Hopfield noticed a striking analogy between the brain and spin glasses, an exotic material much beloved of statistical physicists. This set off a connectionist renaissance that culminated a few years later in the invention of the first algorithms capable of solving the credit-assignment problem, ushering in a new era where machine learning replaced knowledge engineering as the dominant paradigm in AI.

Spin glasses are not actually glasses, although they have some glasslike properties. Rather, they are magnetic materials. Every electron is a tiny magnet by virtue of its spin, which can point "up" or "down." In materials like iron, electrons' spins tend to line up: if an electron with down spin is surrounded by electrons with up spins, it will probably flip to up. When most of the spins in a chunk of iron line up, it turns into a magnet. In ordinary magnets, the strength of interaction between adjacent spins is the same for all pairs, but in a spin glass it can vary; it may even be negative, causing nearby spins to point in opposite directions. The energy of an ordinary magnet is lowest when all its spins align, but in a spin glass, it's not so simple. Indeed, finding the lowest-energy state of a spin glass is an NP-complete problem, meaning that just about every other difficult optimization problem can be reduced to it. Because of this, a spin glass doesn't necessarily settle into its overall lowest energy state; much like rainwater may flow downhill into a lake instead of reaching the ocean, a spin glass may get stuck in a local minimum, a state with lower energy than all the states that can be reached from it by flipping a spin, rather than evolve to the global one.

Hopfield noticed an interesting similarity between spin glasses and neural networks: an electron's spin responds to the behavior of its neighbors much like a neuron does. In the electron's case, it flips up if the weighted sum of the neighbors exceeds a threshold and flips (or stays) down otherwise. Inspired by this, he defined a type of neural network that evolves over time in the same way that a spin glass does and postulated that the network's minimum energy states are its memories. Each such state has a "basin of attraction" of initial states that converge to it, and in this way the network can do pattern recognition: for example, if one of the memories is the pattern of black-and-white pixels formed by the digit nine and the network sees a distorted nine, it will converge to the "ideal" one and thereby recognize it. Suddenly, a vast body of physical theory was applicable to machine learning, and a flood of statistical physicists poured into the field, helping it break out of the local minimum it had been stuck in.

A spin glass is still a very unrealistic model of the brain, though. For one, spin interactions are symmetric, and connections between neurons in the brain are not. Another big issue that Hopfield's model ignored is that real neurons are statistical: they don't deterministically turn on and off as a function of their inputs; rather, as the weighted sum of inputs increases, the neuron becomes more likely to fire, but it's not certain that it will. In 1985, David Ackley, Geoff Hinton, and Terry Sejnowski replaced the deterministic neurons in Hopfield networks with probabilistic ones. A neural network now had a probability distribution over its states, with higher-energy states being exponentially less likely than lower-energy ones. In fact, the probability of finding the network in a particular state was given by the well-known Boltzmann distribution from thermodynamics, so they called their network a Boltzmann machine.

A Boltzmann machine has a mix of sensory and hidden neurons (analogous to, for example, the retina and the brain, respectively). It learns by being alternately awake and asleep, just like humans. While awake, the sensory neurons fire as dictated by the data, and the hidden ones evolve according to the network dynamics and the sensory input. For example, if the network is shown an image of a nine, the neurons corresponding to the black pixels in the image stay on, the others stay off, and the hidden ones fire randomly according to the Boltzmann distribution given those pixel values. During sleep, the machine dreams, leaving both sensory and hidden neurons free to wander. Just before the new day dawns, it compares the statistics of its states during the dream and during yesterday's activities and changes the connection weights so that they match. If two neurons tend to fire together during the day but less so while asleep, the weight of their connection goes up; if it's the opposite, they go down. By doing this day after day, the predicted correlations between sensory neurons evolve until they match the real ones. At this point, the Boltzmann machine has learned a good model of the data and effectively solved the credit-assignment problem.

Geoff Hinton went on to try many variations on Boltzmann machines over the following decades. Hinton, a psychologist turned computer scientist and great-great-grandson of George Boole, the inventor of the logical calculus used in all digital computers, is the world's leading connectionist. He has tried longer and harder to understand how the brain works than anyone else. He tells of coming home from work one day in a state of great excitement, exclaiming "I did it! I've figured out how the brain works!" His daughter replied, "Oh, Dad, not again!" Hinton's latest passion is deep learning, which we'll meet later in this chapter. He was also involved in the development of backpropagation, an even better algorithm than Boltzmann machines for solving the credit-assignment problem that we'll look at next. Boltzmann machines could solve the credit-assignment problem in principle, but in practice learning was very slow and painful, making this approach impractical for most applications. The next breakthrough involved getting rid of another oversimplification that dated all the way back to McCulloch and Pitts.

The most important curve in the world

As far as its neighbors are concerned, a neuron can only be in one of two states: firing or not firing. This misses an important subtlety, however. Action potentials are short lived; the voltage spikes for a small fraction of a second and immediately goes back to its resting state. And a single spike barely registers in the receiving neuron; it takes a train of spikes closely on each other's heels to wake it up. A typical neuron spikes occasionally in the absence of stimulation, spikes more and more frequently as stimulation builds up, and saturates at the fastest spiking rate it can muster, beyond which increased stimulation has no effect. Rather than a logic gate, a neuron is more like a voltage-to-frequency converter. The curve of frequency as a function of voltage looks like this:

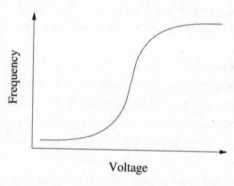

This curve, which looks like an elongated S, is variously known as the logistic, sigmoid, or S curve. Peruse it closely, because it's the most important curve in the world. At first the output increases slowly with the input, so slowly it seems constant. Then it starts to change faster, then very fast, then slower and slower until it becomes almost constant again. The transfer curve of a transistor, which relates its input and output voltages, is also an S curve. So both computers and the brain are filled with S curves. But it doesn't end there. The S curve is the shape of phase transitions of all kinds: the probability of an electron flipping its spin as a function of the applied field, the magnetization of iron, the writing of a bit of memory to a hard disk, an ion channel opening in a cell, ice melting, water evaporating, the inflationary expansion of the early universe, punctuated equilibria in evolution, paradigm shifts in science, the spread of new technologies, white flight from multiethnic neighborhoods, rumors, epidemics, revolutions, the fall of empires, and much more. *The Tipping Point* could equally well (if less appealingly) be entitled *The S Curve*. An earthquake is a phase transition in the relative position of two adjacent tectonic plates. A bump in the night is just the sound of the microscopic tectonic plates in your house's walls shifting, so don't be scared. Joseph Schumpeter said that the economy evolves by cracks and leaps: S curves are the shape of creative destruction. The effect of financial gains and losses on your happiness follows an S curve, so don't sweat the big stuff. The probability that a random logical formula is satisfiable—the quintessential NP-complete problem—undergoes a phase transition from almost 1 to almost 0 as the formula's length increases. Statistical physicists spend their lives studying phase transitions.

In Hemingway's *The Sun Also Rises*, when Mike Campbell is asked how he went bankrupt, he replies, "Two ways. Gradually and then suddenly." The same could be said of Lehman Brothers. That's the essence of an S curve. One of the futurist Paul Saffo's rules of forecasting is: look for the S curves. When you can't get the temperature in the shower just right—first it's too cold, and then it quickly shifts to too hot—blame the S curve. When you make popcorn, watch the S curve's progress: at first nothing happens, then a few kernels pop, then a bunch more, then the bulk of them in a sudden burst of fireworks, then a few more, and then it's ready to eat. Every motion of your muscles follows an S curve: slow, then fast, then slow again. Cartoons gained a new naturalness when the animators at Disney figured this out and started copying it. Your eyes move in S curves, fixating on one thing and then another, along with your consciousness. Mood swings are phase transitions. So are birth, adolescence, falling in love, getting married, getting pregnant, getting a job, losing it, moving to a new town, getting promoted, retiring, and dying. The universe is a vast symphony of phase transitions, from the cosmic to the microscopic, from the mundane to the life changing.

The S curve is not just important as a model in its own right; it's also the jack-of-all-trades of mathematics. If you zoom in on its midsection, it approximates a straight line. Many phenomena we think of as linear are in fact S curves, because nothing can grow without limit.

Because of relativity, and *contra* Newton, acceleration does not increase linearly with force, but follows an S curve centered at zero. So does electric current as a function of voltage in the resistors found in electronic circuits, or in a light bulb (until the filament melts, which is itself another phase transition). If you zoom out from an S curve, it approximates a step function, with the output suddenly changing from zero to one at the threshold. So depending on the input voltages, the same curve represents the workings of a transistor in both digital computers and analog devices like amplifiers and radio tuners. The early part of an S curve is effectively an exponential, and near the saturation point it approximates exponential decay. When someone talks about exponential growth, ask yourself: How soon will it turn into an S curve? When will the population bomb peter out, Moore's law lose steam, or the singularity fail to happen? Differentiate an S curve and you get a bell curve: slow, fast, slow becomes low, high, low. Add a succession of staggered upward and downward S curves, and you get something close to a sine wave. In fact, every function can be closely approximated by a sum of S curves: when the function goes up, you add an S curve; when it goes down, you subtract one. Children's learning is not a steady improvement but an accumulation of S curves. So is technological change. Squint at the New York City skyline and you can see a sum of S curves unfolding across the horizon, each as sharp as a skyscraper's corner.

Most importantly for us, S curves lead to a new solution to the credit-assignment problem. If the universe is a symphony of phase transitions, let's model it with one. That's what the brain does: it tunes the system of phase transitions inside to the one outside. So let's replace the perceptron's step function with an S curve and see what happens.

Climbing mountains in hyperspace

In the perceptron algorithm, the error signal is all or none: you got it either right or wrong. That's not much to go on, particularly if you have a network of many neurons. You may know that the output neuron is wrong (oops, that wasn't your grandmother), but what about some neuron deep inside the brain? What does it even mean for such a neuron to be right or wrong? If the neurons' output is continuous instead of binary, the picture changes. For starters, we now know *how much* the output neuron is wrong by: the difference between it and the desired output. If the neuron should be firing away ("Oh hi, Grandma!") and is firing a little, that's better than if it's not firing at all. More importantly, we can now propagate that error to the hidden neurons: if the output neuron should fire more and neuron A connects to it, then the more A is firing, the more we should strengthen their connection; but if A is inhibited by another neuron B, then B should fire less, and so on. Based on the feedback from all the neurons it's connected to, each neuron decides how much more or less to fire. Based on that and the activity of *its* input neurons, it strengthens or weakens its connections to them. I need to fire more, and neuron B is inhibiting me? Lower its weight. And neuron C is firing away, but its connection to me is weak? Strengthen it. My "customer" neurons, downstream in the network, will tell me how well I'm doing in the next round.

Whenever the learner's "retina" sees a new image, that signal propagates forward through the network until it produces an output. Comparing this output with the desired one yields an error signal, which then propagates back through the layers until it reaches the retina. Based on this returning signal and on the inputs it had received during the forward pass, each neuron adjusts its weights. As the network sees more and more images of your grandmother and other people, the weights gradually converge to values that let it discriminate between the two. Backpropagation, as this algorithm is known, is phenomenally more powerful than the perceptron algorithm. A single neuron could only learn straight lines. Given enough hidden neurons, a multilayer perceptron, as it's called, can represent arbitrarily convoluted frontiers. This makes backpropagation—or simply backprop—the connectionists' master algorithm.

Backprop is an instance of a strategy that is very common in both nature and technology: if you're in a hurry to get to the top of the mountain, climb the steepest slope you can find. The technical term for this is gradient ascent (if you want to get to the top) or gradient descent (if you're looking for the valley bottom). Bacteria can find food by swimming up the concentration gradient of, say, glucose molecules, and they can flee from poisons by swimming down their gradient. All sorts of things, from aircraft wings to antenna arrays, can be optimized by gradient descent. Backprop is an efficient way to do it in a multilayer perceptron: keep tweaking the weights so as to lower the error, and stop when all tweaks fail. With backprop, you don't have to figure out how to tweak each neuron's weights from scratch, which would be too slow; you can do it layer by layer, tweaking each neuron based on how you tweaked the neurons it connects to. If you had to throw out your entire machine-learning toolkit in an emergency save for one tool, gradient descent is probably the one you'd want to hold on to.

So does backprop solve the machine-learning problem? Can we just throw together a big pile of neurons, wait for it to do its magic, and on the way to the bank collect a Nobel Prize for figuring out how the brain works? Alas, life is not that easy. Suppose your network has only one weight, and this is the graph of the error as a function of it:

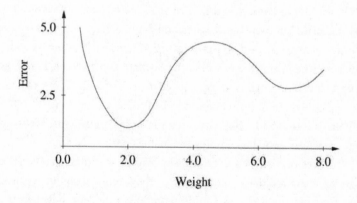

The optimal weight, where the error is lowest, is 2.0. If the network starts out with a weight of 0.75, for example, backprop will get to the optimum in a few steps, like a ball rolling downhill. But if it starts at 5.5, on the other hand, backprop will roll down to 7.0 and remain stuck there. Backprop, with its incremental weight changes, doesn't know how to find the global error minimum, and local ones can be arbitrarily bad, like mistaking your grandmother for a hat. With one weight, you could try every possible value at increments of 0.01 and find the optimum that way. But with thousands of weights, let alone millions or billions, this is not an option because the number of points on the grid goes up exponentially with the number of weights. The global minimum is hidden somewhere in the unfathomable vastness of hyperspace—and good luck finding it.

Imagine you've been kidnapped and left blindfolded somewhere in the Himalayas. Your head is throbbing, and your memory is not too good, either. All you know is you need to get to the top of Mount Everest. What do you do? You take a step forward and nearly slide into a ravine. After catching your breath, you decide to be a bit more systematic. You carefully feel around with your foot until you find the highest point you can and step gingerly to that point. Then you do the same again. Little by little, you get higher and higher. After a while, every step you can take is down, and you stop. That's gradient ascent. If the Himalayas were just Mount Everest, and Everest was a perfect cone, it would work like a charm. But more likely, when you get to a place where every step is down, you're still very far from the top. You're just standing on a foothill somewhere, and you're stuck. That's what happens to backprop, except it climbs mountains in hyperspace instead of 3-D. If your network has a single neuron, just climbing to better weights one step at a time will get you to the top. But with a multilayer perceptron, the landscape is very rugged; good luck finding the highest peak.

This was part of the reason Minsky, Papert, and others couldn't see how to learn multilayer perceptrons. They could imagine replacing step functions by S curves and doing gradient descent, but then they were faced with the problem of local minima of the error. In those days researchers didn't trust computer simulations; they demanded mathematical proof that an algorithm would work, and there's no such proof for backprop. But what we've come to realize is that most of the time a local minimum is fine. The error surface often looks like the quills of a porcupine, with many steep peaks and troughs, but it doesn't really matter if we find the absolute lowest trough; any one will do. Better still, a local minimum may in fact be preferable because it's less likely to prove to have overfit our data than the global one.

Hyperspace is a double-edged sword. On the one hand, the higher dimensional the space, the more room it has for highly convoluted surfaces and local optima. On the other hand, to be stuck in a local optimum you have to be stuck in *every* dimension, so it's more difficult to get stuck in many dimensions than it is in three. In hyperspace there are mountain passes all over the (hyper) place. So, with a little help from a human sherpa, backprop can often find its way to a perfectly good set of weights. It may be only the mystical valley of Shangri-La, not the sea, but why complain if in hyperspace there are millions of Shangri-Las, each with billions of mountain passes leading to it?

Beware of attaching too much meaning to the weights backprop finds, however. Remember that there are probably many very different ones that are just as good. Learning in multilayer perceptrons is a chaotic process in the sense that starting in slightly different places can cause you to wind up at very different solutions. The phenomenon is the same whether the slight difference is in the initial weights or the training data and manifests itself in all powerful learners, not just backprop.

We *could* do away with the problem of local optima by taking out the S curves and just letting each neuron output the weighted sum of its inputs. That would make the error surface very smooth, leaving only one minimum—the global one. The problem, though, is that a linear function of linear functions is still just a linear function, so a network of linear neurons is no better than a single neuron. A linear brain, no matter how large, is dumber than a roundworm. S curves are a nice halfway house between the dumbness of linear functions and the hardness of step functions.

The perceptron's revenge

Backprop was invented in 1986 by David Rumelhart, a psychologist at the University of California, San Diego, with the help of Geoff Hinton and Ronald Williams. Among other things, they showed that backprop can learn XOR, enabling connectionists to thumb their noses at Minsky and Papert. Recall the Nike example: young men and middle-aged women are the most likely buyers of Nike shoes. We can represent this with a network of three neurons: one that fires when it sees a young male, another that fires when it sees a middle-aged female, and another that fires when either of those does. And with backprop we can learn the appropriate weights, resulting in a successful Nike prospect detector. (So there, Marvin.)

In an early demonstration of the power of backprop, Terry Sejnowski and Charles Rosenberg trained a multilayer perceptron to read aloud. Their NETtalk system scanned the text, selected the correct phonemes according to context, and fed them to a speech synthesizer. NETtalk not only generalized accurately to new words, which knowledge-based systems could not, but it learned to speak in a remarkably human-like way. Sejnowski used to mesmerize audiences at research meetings by playing a tape of NETtalk's progress: babbling at first, then starting to make sense, then speaking smoothly with only the occasional error. (You can find samples on YouTube by typing "sejnowski nettalk.")

Neural networks' first big success was in predicting the stock market. Because they could detect small nonlinearities in very noisy data, they beat the linear models then prevalent in finance and their use spread. A typical investment fund would train a separate network for each of a large number of stocks, let the networks pick the most promising ones, and then have human analysts decide which of those to invest in. A few funds, however, went all the way and let the learners themselves buy and sell. Exactly how all these fared is a closely guarded secret, but it's probably not an accident that machine learners keep disappearing into hedge funds at an alarming rate.

Nonlinear models are important far beyond the stock market. Scientists everywhere use linear regression because that's what they know, but more often than not the phenomena they study are nonlinear, and a multilayer perceptron can model them. Linear models are blind to phase transitions; neural networks soak them up like a sponge.

Another notable early success of neural networks was learning to drive a car. Driverless cars first broke into the public consciousness with the DARPA Grand Challenges in 2004 and 2005, but a over a decade earlier, researchers at Carnegie Mellon had already successfully trained a multilayer perceptron to drive a car by detecting the road in video images and appropriately turning the steering wheel. Carnegie Mellon's car managed to drive coast to coast across America with very blurry vision (thirty by thirty-two pixels), a brain smaller than a worm's, and only a few assists from the human copilot. (The project was dubbed "No Hands Across America.") It may not have been the first truly self-driving car, but it did compare favorably with most teenage drivers.

Backprop's applications are now too many to count. As its fame has grown, more of its history has come to light. It turns out that, as is often the case in science, backprop was invented more than once. Yann LeCun in France and others hit on it at around the same time as Rumelhart. A paper on backprop was rejected by the leading AI conference in the early 1980s because, according to the reviewers, Minsky and Papert had already proved that perceptrons don't work. In fact, Rumelhart is credited with inventing backprop by the Columbus test: Columbus was not the first person to discover America, but the last. It turns out that Paul Werbos, a graduate student at Harvard, had proposed a similar algorithm in his PhD thesis in 1974. And in a supreme irony, Arthur Bryson and Yu-Chi Ho, two control theorists, had done the same even earlier: in 1969, the same year that Minsky and Papert published *Perceptrons!* Indeed, the history of machine learning itself shows why we need learning algorithms. If algorithms that automatically find related papers in the scientific literature had existed in 1969, they could have potentially helped avoid decades of wasted time and accelerated who knows what discoveries.

Among the many ironies of the history of the perceptron, perhaps the saddest is that Frank Rosenblatt died in a boating accident in Chesapeake Bay in 1971 and never lived to see the second act of his creation.

A complete model of a cell

A living cell is a quintessential example of a nonlinear system. The cell performs all of its functions by turning raw materials into end products through a complex web of chemical reactions. We can discover the structure of this network using symbolist methods like inverse deduction, as we saw in the last chapter, but to build a complete model of a cell we need to get quantitative, learning the parameters that couple the expression levels of different genes, relate environmental variables to internal ones, and so on. This is difficult because there is no simple linear relationship between these quantities. Rather, the cell maintains its stability through interlocking feedback loops, leading to very complex behavior. Backpropagation is well suited to this problem because of its ability to efficiently learn nonlinear functions. If we had a complete map of the cell's metabolic pathways and enough observations of all the relevant variables, backprop could in principle learn a detailed model of the cell, with a multilayer perceptron to predict each variable as a function of its immediate causes.

For the foreseeable future, however, we'll have only partial knowledge of cells' metabolic networks and be able to observe only a fraction of the variables we'd like to. Learning useful models despite all this missing information, and despite all the inevitable inconsistencies in the information that is available, calls for Bayesian methods, which we'll delve into in Chapter 6. The same goes for making predictions for a particular patient, model in hand: the evidence available is necessarily noisy and incomplete, and Bayesian inference makes the best of it. It helps that, if the goal is to cure cancer, we don't necessarily need to understand all the details of how tumor cells work, only enough to disable them without harming normal cells. In Chapter 6, we'll also see how to orient learning toward the goal while steering clear of the things we don't know and don't need to know.

More immediately, we know we can use inverse deduction to infer the structure of the cell's networks from data and previous knowledge, but there's a combinatorial explosion of ways to apply it, and we need a strategy. Since metabolic networks were designed by evolution, perhaps simulating it in our learning algorithms is the way to go. In the next chapter, we'll see how to do just that.

Deeper into the brain

When backprop first hit the streets, connectionists had visions of quickly learning larger and larger networks until, hardware permitting, they amounted to artificial brains. It didn't turn out that way. Learning networks with one hidden layer was fine, but after that things soon got very difficult. Networks with a few layers worked only if they were carefully designed for the application (character recognition, say). Beyond that, backprop broke down. As we add layers, the error signal becomes more and more diffuse, like a river branching into smaller and smaller tributaries, until we're down to individual raindrops that just don't register. Learning with dozens or hundreds of hidden layers, like the brain, remained a distant dream, and by the mid-1990s, the excitement for multilayer perceptrons had petered out. A hard core of connectionists soldiered on, but by and large the attention of the machine-learning field moved elsewhere. (We'll survey those lands in Chapters 6 and 7.)

Today, however, connectionism is resurgent. We're learning deeper networks than ever before, and they're setting new standards in vision, speech recognition, drug discovery, and other areas. The new field of deep learning is on the front page of the *New York Times*. Look under the hood, and . . . surprise: it's the trusty old backprop engine, still humming. What changed? Nothing much, say the critics: just faster computers and bigger data. To which Hinton and others reply: exactly, we were right all along!

In truth, connectionists have made genuine progress. One of the protagonists of this latest twist in the connectionist roller coaster is an unassuming little device called an autoencoder. An autoencoder is a multilayer perceptron whose output is the same as its input. In goes a picture of your grandmother and out comes—the same picture of your grandmother. At first this seems like a silly idea: What use could such a contraption possibly be? The key is to make the hidden layer much smaller than the input and output layers, so the network can't just learn to copy the input to the hidden layer and the hidden layer to the output, in which case we may as well throw the whole thing out. But if the hidden layer is small, something interesting happens: the network is forced to encode the input in fewer bits, so it can be represented in the hidden layer, and then decode those bits back to full size. It could, for example, learn to encode a million-pixel image of your grandmother as just the seven-character word *grandma*, or some such short code invented by itself, and simultaneously learn to decode "grandma" into an image of dear old granny. So an autoencoder is not unlike a file compression tool, with two important advantages: it figures out how to compress things on its own, and like Hopfield networks, it can turn a noisy, distorted image into a nice clean one.

Autoencoders were known in the 1980s, but they were very hard to learn, even though they had a single hidden layer. Figuring out how to pack a lot of information into the same few bits is a hellishly difficult problem (one code for your grandmother, a slightly different one for your grandfather, another one for Jennifer Aniston, etc). The landscape in hyperspace is just too rugged to get to a good peak; the hidden units need to learn what amounts to too many exclusive-ORs of the inputs. So autoencoders didn't really catch on. The trick that took over a decade to discover was to make the hidden layer larger than the input and output ones. Huh? Actually, that's only half the trick: the other half is to force all but a few of the hidden units to be off at any given time. This still prevents the hidden layer from just copying the input, and—crucially—it makes learning much easier. If we allow different bits to represent different inputs, the inputs no longer have to compete to set the same bits. Also, the network now has many more parameters, so the hyperspace you're in has many more dimensions, and you have many more ways to get out of what would otherwise be local maxima. This is called a sparse autoencoder, and it's a neat trick.

We haven't seen any deep learning yet, though. The next clever idea is to stack sparse autoencoders on top of each other like a club sandwich. The hidden layer of the first autoencoder becomes the input/output layer of the second one, and so on. Because the neurons are nonlinear, each hidden layer learns a more sophisticated representation of the input, building on the previous one. Given a large set of face images, the first autoencoder learns to encode local features like corners and spots, the second uses those to encode facial features like the tip of a nose or the iris of an eye, the third one learns whole noses and eyes, and so on. Finally, the top layer can be a conventional perceptron that learns to recognize your grandmother from the high-level features provided by the layer below it—much easier than using only the crude information provided by a single hidden layer or than trying to backpropagate through all the layers at once. The Google Brain network of *New York Times* fame is a nine-layer sandwich of autoencoders and other ingredients that learns to recognize cats from YouTube videos. At one billion connections, it was at the time the largest network ever learned. It's no surprise that Andrew Ng, one of the project's principals, is also one of the leading proponents of the idea that human intelligence boils down to a single algorithm, and all we need to do is figure it out. Ng, whose affability belies a fierce ambition, believes that stacked sparse autoencoders can take us closer to solving AI than anything that came before.

Stacked autoencoders are not the only kind of deep learner. Another is based on Boltzmann machines, and another—convolutional neural networks—on a model of the visual cortex. Despite their remarkable successes, however, all of these are still a far cry from the brain. The Google network can recognize cat faces seen head on; humans can recognize cats in any pose and even when the face is hard to make out. The Google network is still pretty shallow; only three of its nine layers are autoencoders. A multilayer perceptron is a passable model of the cerebellum, the part of the brain responsible for low-level motor control, but the cortex is another story. It's missing the backward connections needed to propagate errors, for one, and yet it's where the real learning wizardry resides. In his book *On Intelligence*, Jeff Hawkins advocated designing algorithms closely based on the organization of the cortex, but so far none of these algorithms can compete with today's deep networks.

This may change as our understanding of the brain improves. Inspired by the human genome project, the new field of connectomics seeks to map every synapse in the brain. The European Union is investing a billion euros to build a soup-to-nuts model of it. America's BRAIN initiative, with $100 million in funding in 2014 alone, has similar aims. Nevertheless, symbolists are very skeptical of this path to the Master Algorithm. Even if we can image the whole brain at the level of individual synapses, we (ironically) need better machine-learning algorithms to turn those images into wiring diagrams; doing it by hand is out of the question. Worse than that, even if we had a complete map of the brain, we would still be at a loss to figure out what it does. The nervous system of the *C. elegans* worm consists of only 302 neurons and was completely mapped in 1986, but we still have only a fragmentary understanding of what it does. We need higher-level concepts to make sense of the morass of low-level details, weeding out the ones that are specific to wetware or just quirks of evolution. We don't build airplanes by reverse engineering feathers, and airplanes don't flap their wings. Rather, airplane designs are based on the principles of aerodynamics, which all flying objects must obey. We still do not understand those analogous principles of thought.

Perhaps connectomics is overkill. Some connectionists have been overheard claiming that backprop is the Master Algorithm and we just need to scale it up. But symbolists pour scorn on this notion. They point to a long list of things that humans can do but neural networks can't. Take commonsense reasoning. It involves combining pieces of information that may have never been seen together before. Did Mary eat a shoe for lunch? No, because Mary is a person, people only eat edible things, and shoes are not edible. Symbolic systems have no trouble with this—they just chain the relevant rules—but multilayer perceptrons can't do it; once they're done learning, they just compute the same fixed function over and over again. Neural networks are not compositional, and compositionality is a big part of human cognition. Another big issue is that humans—and symbolic models like sets of rules and decision trees—can explain their reasoning, while neural networks are big piles of numbers that no one can understand.

But if humans have all these abilities that their brains didn't learn by tweaking synapses, where did they come from? Unless you believe in magic, the answer must be evolution. If you're a connectionism skeptic and you have the courage of your convictions, it behooves you to figure out how evolution learned everything a baby knows at birth—and the more you think is innate, the taller the order. But if you can figure it out and program a computer to do it, it would be churlish to deny that you've invented at least one version of the Master Algorithm.

CHAPTER FIVE

Evolution: Nature's Learning Algorithm

Robotic Park is a massive robot factory surrounded by ten thousand square miles of jungle, urban and otherwise. Ringing that jungle is the tallest, thickest wall ever built, bristling with sentry posts, searchlights, and gun turrets. The wall has two purposes: to keep trespassers out and the park's inhabitants—millions of robots battling for survival and control of the factory—within. The winning robots get to spawn, their reproduction accomplished by programming the banks of 3-D printers inside. Step-by-step, the robots become smarter, faster—and deadlier. Robotic Park is run by the US Army, and its purpose is to evolve the ultimate soldier.

Robotic Park doesn't exist yet, but it may someday. I suggested it as a thought experiment at a DARPA workshop a few years ago, and one of the military brass present said matter-of-factly, "That's feasible." His willingness might seem less startling if you consider that the army already runs a full-blown mockup of an Afghan village in the California desert, complete with villagers, for training its troops, and a few billion dollars would be a small price to pay for the ultimate soldier.

The first steps toward Robotic Park have already been taken. Inside Hod Lipson's Creative Machines Lab at Cornell University, fantastically shaped robots are learning to crawl and fly, probably even as you read this. One looks like a slithering tower of rubber bricks, another like a helicopter with dragonfly wings, yet another like a shape-shifting Tinkertoy. These robots were not designed by any human engineer but created by evolution, the same process that gave rise to the diversity of life on Earth. Although the robots initially evolve inside a computer simulation, once they look proficient enough to make it in the real world, solid versions are automatically fabricated by 3-D printing. These are not yet ready to take over the world, but they've come a long way from the primordial soup of simulated parts they started with.

The algorithm that evolved these robots was invented by Charles Darwin in the nineteenth century. He didn't think of it as an algorithm at the time, partly because a key subroutine was still missing. Once James Watson and Francis Crick provided it in 1953, the stage was set for the second coming of evolution: *in silico* instead of *in vivo*, and a billion times faster. Its prophet was a ruddy-faced, perpetually grinning midwesterner by the name of John Holland.

Darwin's algorithm

Like many other early machine-learning researchers, Holland started out working on neural networks, but his interests took a different turn when, while a graduate student at the University of Michigan, he read Ronald Fisher's classic treatise *The Genetical Theory of Natural Selection*. In it, Fisher, who was also the founder of modern statistics, formulated the first mathematical theory of evolution. Brilliant as it was, Holland felt that Fisher's theory left out the essence of evolution. Fisher considered each gene in isolation, but an organism's fitness is a complex function of all its genes. If genes are independent, the relative frequencies of their variants rapidly converge to the maximum fitness point and remain in equilibrium thereafter. But if genes interact, evolution—the search for maximum fitness—is vastly more complex. With one thousand genes, each with two variants, the genome has 2^{1000} possible states, and no planet in the universe is remotely large or ancient enough to have tried them all out. Yet on Earth evolution has managed to come up with some remarkably fit organisms, and Darwin's theory of natural selection explains how, at least qualitatively. Holland decided to turn it into an algorithm.

But first he had to graduate. Prudently, he picked a more conservative topic for his dissertation—Boolean circuits with cycles—and in 1959 he earned the world's first PhD in computer science. His PhD advisor, Arthur Burks, nevertheless encouraged Holland's interest in evolutionary computation and was instrumental in getting him a faculty job at Michigan and shielding him from senior colleagues who didn't think that stuff was computer science. Burks himself was so open-minded because he had been a close collaborator of John von Neumann, who had proved the possibility of self-reproducing machines. Indeed, it had fallen to him to complete the work when von Neumann died of cancer in 1957. That von Neumann could prove that such machines are possible was quite remarkable, given the primitive state of genetics and computer science at the time. But his automaton just made exact copies of itself; evolving automata had to wait for Holland.

The key input to a genetic algorithm, as Holland's creation came to be known, is a fitness function. Given a candidate program and some purpose it is meant to fill, the fitness function assigns the program a numeric score reflecting how well it fits the purpose. In natural selection, it's questionable whether fitness can be interpreted this way: while the fitness of a wing for flight makes intuitive sense, evolution as a whole has no known purpose. Nevertheless, in machine learning having something like a fitness function is a no-brainer. If we need a program that can diagnose a patient, one that correctly diagnoses 60 percent of the patients in our database is better than one that only gets it right 55 percent of the time, and thus a possible fitness function is the fraction of correctly diagnosed cases.

In this regard, genetic algorithms are a lot like selective breeding. Darwin opened *The Origin of Species* with a discussion of it, as a stepping-stone to the more difficult concept of natural selection. All the domesticated plants and animals we take for granted today are the result of selecting and mating, generation after generation, the organisms that best served our purposes: the corn with the largest corncobs, the sweetest fruit trees, the shaggiest sheep, the hardiest horses. Genetic algorithms do the same, except they breed programs instead of living creatures, and a generation is a few seconds of computer time instead of a creature's lifetime.

The fitness function encapsulates the human's role in the process. But the more subtle part is nature's. Starting with a population of not-very-fit individuals—possibly completely random ones—the genetic algorithm has to come up with variations that can then be selected according to fitness. How does nature do that? Darwin didn't know. This is where the genetic part of the algorithm comes in. In the same way that DNA encodes an organism as a sequence of base pairs, we can encode a program as a string of bits. Instead of 0 and 1, the DNA alphabet has four characters—the four bases adenine, thymine, cytosine, and guanine—but that's a superficial difference. Variations, whether in DNA sequences or bit strings, can be generated in several ways. The simplest approach is point mutation, flipping a random bit in the string or changing a single base in a stretch of DNA. But for Holland, the real power of genetic algorithms lay in something more complicated: sex.

Stripped down to its bare essentials (no giggles, please), sexual reproduction consists of swapping material between chromosomes from the mother and father, a process called crossing over. This produces two new chromosomes, one of which consists of the mother's chromosome up to the crossover point and the father's thereafter, and the other one is the opposite:

PARENTS   OFFSPRING

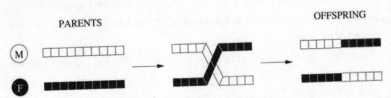

A genetic algorithm works by mimicking this process. In each generation, it mates the fittest individuals, producing two offspring from each pair of parents by crossing over their bit strings at a random point. After applying point mutations to the new strings, it lets them loose in its virtual world. Each one returns with a fitness score, and the process repeats. Each generation is fitter than the previous one, and the process terminates when the desired fitness is reached or time runs out.

For example, suppose we want to evolve a rule for filtering spam. If ten thousand different words appear in the training data, each candidate rule can be represented by a string of twenty thousand bits, two for each word. The first bit corresponding to the word *free* is one if e-mails containing *free* are allowed to match the rule, and zero if they're not. The second bit is the opposite: one if e-mails *not* containing *free* are allowed to match, and zero if they're not. So if both bits are one, e-mails are allowed to match the rule regardless of whether they contain *free*, and the rule effectively has no condition on that word. On the other hand, if both bits are zero, no e-mails match the rule, since one or the other bit always fails, and all e-mails get through the filter (yikes). Overall, an e-mail matches a rule only if its entire pattern of present and absent words is allowed by the rule. A rule's fitness is, say, the percentage of e-mails it classifies correctly. Starting from a population of random strings, each representing a rule with random conditions, the genetic algorithm can now evolve better and better rules by repeatedly crossing over and mutating the fittest strings in each generation. For example, if the current population includes the rules *If the e-mail contains the word* free *then it's spam* and *If the e-mail contains the word* easy *then it's spam,* crossing them over will yield the probably fitter rule *If the e-mail contains* free *and* easy *then it's spam,* provided the crossover point does not fall between the two bits corresponding to one of those words. It will also yield the rule *All e-mail is spam,* which results from dropping both conditions, but that rule is unlikely to have much progeny in the next generation.

Since our goal is to produce the best spam filter we can, as opposed to faithfully simulating real natural selection, we can cheat liberally by modifying the algorithm to fit our needs. One way in which genetic algorithms routinely cheat is by allowing immortality. (Too bad we can't do that in real life.) That way, a highly fit individual doesn't simply compete to reproduce within its own generation, but also with its children, and then its grandchildren, great-grandchildren, and so on, as long as it remains one of the fittest individuals in the population. In contrast, in the real world the best a highly fit individual can do is pass on half its genes to many children, each of which will probably be less fit because of the genes it inherited from its other parent. Immortality avoids this backsliding and with any luck, lets the algorithm reach the desired fitness sooner. Of course, since the fittest humans in history as measured by number of descendants are the likes of Genghis Khan—ancestor to one in two hundred men alive today—perhaps it's not so bad that in real life immortality is *verboten*.

If we want to evolve a whole set of spam-filtering rules, not just one, we can represent a candidate set of n rules by a string of $n \times 20,000$ bits (20,000 for each rule, assuming ten thousand different words in the data, as before). Rules containing 00 for some word effectively disappear from the rule set, since they don't match any e-mails, as we saw before. If an e-mail matches any rule in the set, it's classified as spam; otherwise it's legit. We can still let fitness be the percentage of correctly classified e-mails, but to combat overfitting, we'll probably want to subtract from it a penalty proportional to the total number of active conditions in the rule set.

We can get even fancier by allowing rules for intermediate concepts to evolve, and then chaining these rules at performance time. For example, we could evolve the rules *If the e-mail contains the word* loan *then it's a scam* and *If the e-mail is a scam then it's spam*. Since a rule's consequent is no longer always *spam*, this requires introducing additional bits in rule strings to represent their consequents. Of course, the computer doesn't literally use the word *scam*; it just comes up with some arbitrary bit string to represent the concept, but that's good enough for our purposes. Sets of rules like this, which Holland called classifier systems, are one of the workhorses of the machine-learning tribe he founded: the evolutionaries. Like multilayer perceptrons, classifier systems face the credit-assignment problem—what is the fitness of rules for intermediate concepts?—and Holland devised the so-called bucket brigade algorithm to solve it. Nevertheless, classifier systems are much less widely used than multilayer perceptrons.

Compared to the simple model in Fisher's book, genetic algorithms are quite a leap forward. Darwin lamented his lack of mathematical ability, but if he had lived a century later he probably would have yearned for programming prowess instead. Indeed, capturing natural selection by a set of equations is extremely difficult, but expressing it as an algorithm is another matter, and can shed light on many otherwise vexing questions. Why do species appear suddenly in the fossil record? Where's the evidence that they evolved gradually from earlier species? In 1972, Niles Eldredge and Stephen Jay Gould proposed that evolution consists of a series of "punctuated equilibria," alternating long periods of stasis with short bursts of rapid change, like the Cambrian explosion. This sparked a heated debate, with critics of the theory nicknaming it "evolution by jerks" and Eldredge and Gould retorting that gradualism is "evolution by creeps." Experience with genetic algorithms lends support to the jerks. If you run a genetic algorithm for one hundred thousand generations and observe the population at one-thousand-generation intervals, the graph of fitness against time will probably look like an uneven staircase, with sudden improvements followed by flat periods that tend to become longer over time. It's also not hard to see why. Once the algorithm reaches a local maximum of fitness—a peak in the fitness landscape—it will stay there for a long time until a lucky mutation or crossover lands an individual on the slope to a higher peak, at which point that individual will multiply and climb up the slope with each passing generation. And the higher the current peak, the longer before that happens. Of course, natural evolution is more complicated than this: for one, the environment may change, either physically or because other organisms have themselves evolved, and an organism that was on a fitness peak may suddenly find itself under pressure to evolve again. So, while helpful, current genetic algorithms are far from the end of the story.

The exploration-exploitation dilemma

Notice how much genetic algorithms differ from multilayer perceptrons. Backprop entertains a single hypothesis at any given time, and the hypothesis changes gradually until it settles into a local optimum. Genetic algorithms consider an entire population of hypotheses at each step, and these can make big jumps from one generation to the next, thanks to crossover. Backprop proceeds deterministically after setting the initial weights to small random values. Genetic algorithms, in contrast, are full of random choices: which hypotheses to keep alive and cross over (with fitter hypotheses being more likely candidates), where to cross two strings, which bits to mutate. Backprop learns weights for a predefined network architecture; denser networks are more flexible but also harder to learn. Genetic algorithms make no a priori assumptions about the structures they will learn, other than their general form.

Because of all this, genetic algorithms are much less likely than backprop to get stuck in a local optimum and in principle better able to come up with something truly new. But they are also much more difficult to analyze. How do we know a genetic algorithm will get somewhere meaningful instead of randomly walking around like the proverbial drunkard? The key is to think in terms of building blocks. Every subset of a string's bits potentially encodes a useful building block, and when we cross over two strings, those building blocks come together into a larger one, which in turn becomes grist for the mill. Holland likes to use police sketches to illustrate the power of building blocks. In the days before computers, a police artist could quickly put together a portrait of a suspect from eyewitness interviews by selecting a mouth from a set of paper strips depicting typical mouth shapes and doing the same for the eyes, nose, chin, and so on. With only ten building blocks and ten options for each, this system would allow for ten billion different faces, more than there are people on Earth.

In machine learning, as elsewhere in computer science, there's nothing better than getting such a combinatorial explosion to work for you instead of against you. What's clever about genetic algorithms is that each string implicitly contains an exponential number of building blocks, known as schemas, and so the search is a lot more efficient than it seems. This is because every subset of the string's bits is a schema, representing some potentially fit combination of properties, and a string has an exponential number of subsets. We can represent a schema by replacing the bits in the string that aren't part of it with *. For example, the string 110 contains the schemas ***, **0, *1*, 1**, *10, 11*, 1*0, and 110. We get a different schema for every different choice of bits to include; since we have two choices for each bit (include/don't include), we have 2^n schemas. Conversely, a particular schema may be represented in many different strings in a population, and is implicitly evaluated every time they are. Suppose that a hypothesis's probability of surviving into the next generation is proportional to its fitness. Holland showed that, in this case, the fitter a schema's representatives in one generation are compared to the average, the more of them we can expect to see in the next generation. So, while the genetic algorithm explicitly manipulates strings, it implicitly searches the much larger space of schemas. Over time, fitter schemas come to dominate the population, and so unlike the drunkard, the genetic algorithm finds its way home.

One of the most important problems in machine learning—and life—is the exploration-exploitation dilemma. If you've found something that works, should you just keep doing it? Or is it better to try new things, knowing it could be a waste of time but also might lead to a better solution? Would you rather be a cowboy or a farmer? Start a company or run an existing one? Go steady or play the field? A midlife crisis is the yearning to explore after many years spent exploiting. On an impulse, you fly to Vegas, ready to gamble away your life's savings on the chance of becoming a millionaire. You enter the first casino and face a row of slot machines. The one to play is the one that gives you the best payoff on average, but you don't know which that is. You have to try each one enough times to figure it out. But if you do this for too long, you waste your money on losing machines. Conversely, if you jump the gun and pick a machine that looked good by chance on the first few turns but is in fact not the best one, you waste your money playing it for the rest of the night. That's the exploration-exploitation dilemma. Each time you play, you have to choose between repeating the best move you've found so far, which gives you the best payoff, or trying other moves, which gather information that may lead to even better payoffs. With two slot machines, Holland showed that the optimal strategy is to flip a biased coin each time, where the coin becomes exponentially more biased as you go along. (Don't sue me if it doesn't work for you, though. Remember the house always wins in the end.) The better a slot machine looks, the more you should play it, but never completely give up on the other one, in case it turns out to be the best one after all.

A genetic algorithm is like the ringleader of a group of gamblers, playing slot machines in every casino in town at the same time. Two schemas compete with each other if they include the same bits and differ in at least one of them, like *10 and *11, and n competing schemas are like n slot machines. Every set of competing schemas is a casino, and the genetic algorithm simultaneously figures out the winning machine in every casino, following the optimal strategy of playing the better-seeming machines with exponentially increasing frequency. Pretty smart.

In *The Hitchhiker's Guide to the Galaxy*, an alien race builds a massive supercomputer to answer the ultimate question, and after a long time the computer spits out "42." But the computer also points out that the aliens don't know what the question is, so they build an even bigger computer to figure that out. This computer—otherwise known as planet Earth—is unfortunately destroyed to make way for a space freeway minutes before finishing its multimillion-year computation. We can only guess at the question now, but perhaps it was: Which slot machine should you play?

Survival of the fittest programs

For the first few decades, the genetic algorithms community consisted mainly of John Holland, his students, and their students. Circa 1983, the biggest problem genetic algorithms had been able to solve was learning to control gas pipeline systems. But then, at around the same time neural networks were making their comeback, interest in evolutionary computation took off. The first international conference on genetic algorithms was held in Pittsburgh in 1985, and a Cambrian explosion of genetic algorithm variants was under way. Some of these tried to model evolution more closely—the basic genetic algorithm was only a very crude approximation, after all—and others radiated in very different directions, crossing over evolutionary ideas with computer science concepts that would have bemused Darwin.

One of Holland's more remarkable students was John Koza. In 1987, while flying back to California from a conference in Italy, he had a lightbulb moment. Instead of evolving comparatively simple things like *If . . . then . . .* rules and gas pipeline controllers, why not evolve full-blown computer programs? And if that's the goal, why stick with bit strings as the representation? A program is really a tree of subroutine calls, so better to directly cross over those subtrees than to shoehorn them into bit strings and run the risk of destroying perfectly good subroutines when you cross them over at a random point.

For example, suppose you want to evolve a program to compute the duration of a planet's year, T, from its average distance to the sun, D. According to Kepler's third law, T is the square root of D cubed, times a constant C that depends on the units you use for time and distance. A genetic algorithm should be able to discover this by looking at Tycho Brahe's data on planetary motions like Kepler did. In Koza's approach, D and C are the leaves of a program tree, and the operations that combine them, like multiplication and taking the square root, are the internal nodes. The following program tree correctly computes T:

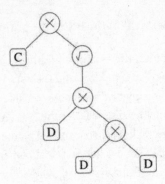

In genetic programming, as Koza called his method, we cross over two program trees by randomly swapping two of their subtrees. For example, crossing over these two trees at the highlighted nodes yields the correct program for computing *T* as one of the children:

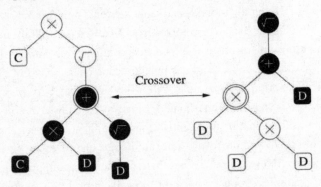

We can measure a program's fitness (or lack thereof) by the distance between its output and the correct one on the training data. For example, if the program says an Earth year is three hundred days, that would subtract sixty-five points from its fitness. Starting with a population of random program trees, genetic programming uses crossover, mutation, and survival to gradually evolve better programs until it's satisfied.

Of course, computing the length of a planet's year is a very simple problem, involving only multiplication and square roots. In general, program trees can include the full range of programming constructs, such as *If . . . then . . .* statements, loops, and recursion. A more illustrative example of what genetic programming can do is figuring out the sequence of actions a robot needs to perform to achieve some goal. Suppose I ask my officebot to bring me a stapler from the closet down the hall. The robot has a large set of behaviors available to it, such as moving down a hallway, opening a door, picking up an object, and so on. Each of these can in turn be composed of various sub-behaviors: move the robot's hand toward the object, or grasp it at various possible points, for example. Each behavior may be executed or not depending on the results of previous behaviors, may need to be repeated some number of times, and so on. The challenge is to assemble the right structure of behaviors and sub-behaviors, together with the parameters for each, such as how far to move the hand. Starting with the robot's "atomic" behaviors and their allowed combinations, genetic programming can assemble a complex behavior that accomplishes the desired goal. A number of researchers have evolved strategies for robot soccer players in this way.

One consequence of crossing over program trees instead of bit strings is that the resulting programs can have any size, making the learning more flexible. The overall tendency is for bloat, however, with larger and larger trees growing as evolution goes on longer (also known as "survival of the fattest"). Evolutionaries can take comfort from the fact that human-written programs are no different (Microsoft Windows: forty-five million lines of code and counting), and that human-made code doesn't allow a solution as simple as adding a complexity penalty to the fitness function.

Genetic programming's first success, in 1995, was in designing electronic circuits. Starting with a pile of electronic components such as transistors, resistors, and capacitors, Koza's system reinvented a previously patented design for a low-pass filter, a circuit that can be used for things like enhancing the bass on a dance-music track. Since then he's made a sport of reinventing patented devices, turning them out by the dozen. The next milestone came in 2005, when the US Patent and Trademark Office awarded a patent to a genetically designed factory optimization system. If the Turing test had been to fool a patent examiner instead of a conversationalist, then January 25, 2005, would have been a date for the history books.

Koza's confidence stands out even in a field not known for its shrinking violets. He sees genetic programming as an invention machine, a silicon Edison for the twenty-first century. He and other evolutionaries believe it can learn any program, making it their entry in the Master Algorithm sweepstakes. In 2004, they instituted the annual Humie Awards to recognize "human-competitive" genetic creations; thirty-nine have been awarded to date.

What is sex for?

Despite their successes, and the insights they've provided on issues like gradualism versus punctuated equilibria, genetic algorithms have left one great mystery unsolved: the role of sex in evolution. Evolutionaries set great store by crossover, but members of the other tribes think it's not worth the trouble. None of Holland's theoretical results show that crossover actually helps; mutation suffices to exponentially increase the frequency of the fittest schemas in the population over time. And the "building blocks" intuition is appealing but quickly runs into trouble, even when genetic programming is used. As larger blocks evolve, crossover also becomes increasingly likely to break them up. Also, once a highly fit individual appears, its descendants tend to quickly take over the population, crowding out potentially better schemas that were trapped in overall less fit individuals. This effectively reduces the search to variations of the fitness champ. Researchers have come up with a number of schemes for preserving diversity in the population, but the results so far are inconclusive. Engineers certainly use building blocks extensively, but combining them involves, well, a lot of engineering; it's not just a matter of throwing them together any old way, and it's not clear crossover can do the trick.

Eliminating sex would leave evolutionaries with only mutation to power their engine. If the size of the population is substantially larger than the number of genes, chances are that every point mutation is represented in it, and the search becomes a type of hill climbing: try all possible one-step variations, pick the best one, and repeat. (Or pick several of the best variations, in which case it's called beam search.) Symbolists, in particular, use this all the time to learn sets of rules, although they don't think of it as a form of evolution. To avoid getting trapped in local maxima, hill climbing can be enhanced with randomness (make a downhill move with some probability) and random restarts (after a while, jump to a random state and continue from there). Doing this is enough to find good solutions to problems; whether the benefit of adding crossover to it justifies the extra computational cost remains an open question.

No one is sure why sex is pervasive in nature, either. Several theories have been proposed, but none is widely accepted. The leader of the pack is the Red Queen hypothesis, popularized by Matt Ridley in the eponymous book. As the Red Queen said to Alice in *Through the Looking Glass*, "It takes all the running you can do, to keep in the same place." In this view, organisms are in a perpetual arms race with parasites, and sex helps keep the population varied, so that no single germ can infect all of it. If this is the answer, then sex is irrelevant to machine learning, at least until learned programs have to vie with computer viruses for processor time and memory. (Intriguingly, Danny Hillis claims that deliberately introducing coevolving parasites into a genetic algorithm can help it escape local maxima by gradually ratcheting up the difficulty, but no one has followed up on this yet.) Christos Papadimitriou and colleagues have shown that sex optimizes not fitness but what they call mixability: a gene's ability to do well on average when combined with other genes. This can be useful when the fitness function is either not known or not constant, as in natural selection, but in machine learning and optimization, hill climbing tends to do better.

The problems for genetic programming do not end there. Indeed, even its successes might not be as genetic as evolutionaries would like. Take circuit design, which was genetic programming's emblematic success. As a rule, even relatively simple designs require an enormous amount of search, and it's not clear how much the results owe to brute force rather than genetic smarts. To address the growing chorus of critics, Koza included in his 1992 book *Genetic Programming* experiments showing that genetic programming beat randomly generating candidates on Boolean circuit synthesis problems, but the margin of victory was small. Then, at the 1995 International Conference on Machine Learning (ICML) in Lake Tahoe, California, Kevin Lang published a paper showing that hill climbing beat genetic programming on the same problems, often by a large margin. Koza and other evolutionaries had repeatedly tried to publish papers in ICML, a leading venue in the field, but to their increasing frustration they kept being rejected due to insufficient empirical validation. Already frustrated with his papers being rejected, seeing Lang's paper made Koza blow his top. On short order, he produced a twenty-three-page paper in two-column ICML format refuting Lang's conclusions and accusing the ICML reviewers of scientific misconduct. He then placed a copy on every seat in the conference auditorium. Depending on your point of view, either Lang's paper or Koza's response was the last straw; regardless, the Tahoe incident marked the final divorce between the evolutionaries and the rest of the machine-learning community, with the evolutionaries moving out of the house. Genetic programmers started their own conference, which merged with the genetic algorithms conference to form GECCO, the Genetic and Evolutionary Computing Conference. For its part, the machine-learning mainstream largely forgot them. A sad *dénouement*, but not the first time in history that sex is to blame for a breakup.

Sex may not have succeeded in machine learning, but as a consolation, it has played a prominent role in the evolution of technology in other ways. Pornography was the unacknowledged "killer app" of the World Wide Web, not to mention the printing press, photography, and video before it. The vibrator was the first handheld electrical device, predating the cell phone by a century. Scooters took off in postwar Europe, particularly Italy, because they let young couples get away from their families. Facilitating dating was surely one of the "killer apps" of fire when *Homo erectus* discovered it a million years ago; and equally surely, a key driver of increasing realism in humanlike robots will be the sexbot industry. Sex just seems to be the end, rather than the means, of technological evolution.

Nurturing nature

Evolutionaries and connectionists have something important in common: they both design learning algorithms inspired by nature. But then they part ways. Evolutionaries focus on learning structure; to them, fine-tuning an evolved structure by optimizing parameters is of secondary importance. In contrast, connectionists prefer to take a simple, hand-coded structure with lots of connections and let weight learning do all the work. This is machine learning's version of the nature versus nurture controversy, and there are good arguments on both sides.

On the one hand, evolution has produced many amazing things, none more amazing than you. With or without crossover, evolving structure is an essential part of the Master Algorithm. The brain can learn anything, but it can't evolve a brain. If we thoroughly understood its architecture, we could just implement it in hardware, but we're very far from that; getting an assist from computer-simulated evolution is a no-brainer. What's more, we also want to evolve the brains of robots, systems with arbitrary sensors, and super-AIs. There's no reason to stick with the design of the human brain if there are better ones for those tasks. On the other hand, evolution is excruciatingly slow. The entire life of an organism yields only one piece of information about its genome: its fitness, reflected in the organism's number of offspring. That's a colossal waste of information, which neural learning avoids by acquiring the information at the point of use (so to speak). As connectionists like Geoff Hinton like to point out, there's no advantage to carrying around in the genome information that we can readily acquire from the senses. When a newborn opens his eyes, the visual world comes flooding in; the brain just has to organize it. What does need to be specified in the genome, however, is the architecture of the machine that does the organizing.

As in the nature versus nurture debate, neither side has the whole answer; the key is figuring out how to combine the two. The Master

Algorithm is neither genetic programming nor backprop, but it has to include the key elements of both: structure learning and weight learning. In the conventional view, nature does its part first—evolving a brain—and then nurture takes it from there, filling the brain with information. We can easily reproduce this in learning algorithms. First, learn the structure of the network, using (for example) hill climbing to decide which neurons connect to which: try adding each possible new connection to the network, keep the one that most improves performance, and repeat. Then learn the connection weights using backprop, and your brand-new brain is ready to use.

But now there's an important subtlety, in both natural and artificial evolution. We need to learn weights for every candidate structure along the way, not just the final one, in order to see how well it does in the struggle for life (in the natural case) or on the training data (in the artificial case). The structure we want to select at each step is the one that does best after learning weights, not before. So in reality, nature does not come before nurture; rather, they alternate, with each round of "nurture" learning setting the stage for the next round of "nature" learning and vice versa. Nature evolves for the nurture it gets. The evolutionary growth of the cortex's associative areas builds on neural learning in the sensory areas, without which it would be useless. Goslings follow their mother around (evolved behavior) but that requires recognizing her (learned ability). If you're the first thing they see when they hatch, they'll follow you instead, as Konrad Lorenz memorably showed. The newborn brain already encodes features of the environment but not explicitly; rather, evolution optimized it to extract those features from the expected input. Likewise, in an algorithm that iteratively learns both structure and weights, each new structure is implicitly a function of the weights learned in previous rounds.

Of all the possible genomes, very few correspond to viable organisms. The typical fitness landscape thus consists of vast flatlands with occasional sharp peaks, making evolution very hard. If you start out blindfolded in Kansas, you have no idea which way the Rockies lie, and you'll wander around for a long time before you bump into their foothills and start climbing. But if you combine evolution with neural learning, something interesting happens. If you're on flat ground, but not too far from the foothills, neural learning can get you there, and the closer you are to the foothills, the more likely it will. It's like being able to scan the horizon: it won't help you in Wichita, but in Denver you'll see the Rockies in the distance and head that way. Denver now looks a lot fitter than it did when you were blindfolded. The net effect is to widen the fitness peaks, making it possible for you to find your way to them from previously very tough places, like point A in this graph:

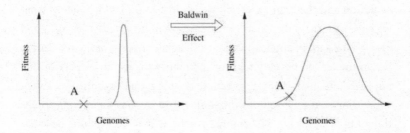

In biology, this is called the Baldwin effect, after J. M. Baldwin, who proposed it in 1896. In Baldwinian evolution, behaviors that are first learned later become genetically hardwired. If dog-like mammals can learn to swim, they have a better chance to evolve into seals—as they did—than if they drown. Thus individual learning can influence evolution without recourse to Lamarckism. Geoff Hinton and Steven Nowlan demonstrated the Baldwin effect in machine learning by using genetic algorithms to evolve neural network structure and observing that fitness increased over time only when individual learning was allowed.

He who learns fastest wins

Evolution searches for good structures, and neural learning fills them in: this combination is the easiest of the steps we'll take toward the Master Algorithm. This may come as a surprise to anyone familiar with the never-ending twists and turns of the nature versus nurture controversy,

2,500 years old and still going strong. Seeing life through the eyes of a computer clarifies a lot of things, however. "Nature" for a computer is the program it runs, and "nurture" is the data it gets. The question of which one is more important is clearly absurd; there's no output without both program and data, and it's not like the output is, say, 60 percent caused by the program and 40 percent by the data. That's the kind of linear thinking that a familiarity with machine learning immunizes you against.

On the other hand, you may be wondering why we're not done at this point. Surely if we've combined nature's two master algorithms, evolution and the brain, that's all we could ask for. Unfortunately, what we have so far is only a very crude cartoon of how nature learns, good enough for a lot of applications but still a pale shadow of the real thing. For example, the development of the embryo is a crucial part of life, but there's no analog of it in machine learning: the "organism" is a very straightforward function of the genome, and we may be missing something important there. But another reason is that we wouldn't be satisfied even if we had completely figured out how nature learns. For one thing, it's too slow. Evolution takes billions of years to learn, and the brain takes a lifetime. Culture is better: I can distill a lifetime of learning into a book, and you can read it in a few hours. But learning algorithms should be able to learn in minutes or seconds. He who learns fastest wins, whether it's the Baldwin effect speeding up evolution, verbal communication speeding up human learning, or computers discovering patterns at the speed of light. Machine learning is the latest chapter in the arms race of life on Earth, and swifter hardware is only half the equation. The other half is smarter software.

Most of all, the goal of machine learning is to find the best possible learning algorithm, by any means available, and evolution and the brain are unlikely to provide it. The products of evolution have many obvious faults. For example, the mammalian optic nerve attaches to the front of the retina instead of the back, causing an unnecessary—and egregious—blind spot right next to the fovea, the area of sharpest vision.

The molecular biology of living cells is such a mess that molecular biologists often quip that only people who don't know any of it could believe in intelligent design. The architecture of the brain may well have similar faults—the brain has many constraints that computers don't, like very limited short-term memory—and there's no reason to stay within them. Moreover, we know of many situations where humans seem to consistently do the wrong thing, as Daniel Kahneman illustrates at length in his book *Thinking, Fast and Slow*.

In contrast to the connectionists and evolutionaries, symbolists and Bayesians do not believe in emulating nature. Rather, they want to figure out from first principles what learners should do—and that includes us humans. If we want to learn to diagnose cancer, for example, it's not enough to say "this is how nature learns; let's do the same." There's too much at stake. Errors cost lives. Doctors should diagnose in the most foolproof way they can, with methods similar to those mathematicians use to prove theorems, or as close to that as they can manage, given that it's seldom possible to be that rigorous. They need to weigh the evidence to minimize the chances of a wrong diagnosis; or more precisely, so that the costlier an error is, the less likely they are to make it. (For example, failing to find a tumor that's really there is potentially much worse than inferring one that isn't.) They need to make *optimal* decisions, not just decisions that seem good.

This is an instance of a tension that runs throughout much of science and philosophy: the split between descriptive and normative theories, between "this is how it is" and "this is how it should be." Symbolists and Bayesians like to point out, however, that figuring out how we should learn can also help us to understand how we do learn because the two are presumably not entirely unrelated—far from it. In particular, behaviors that are important for survival and have had a long time to evolve should not be far from optimal. We're not very good at answering written questions about probabilities, but we are very good at instantly choosing hand and arm movements to hit a target. Many psychologists have used symbolist or Bayesian models to explain aspects of human behavior. Symbolists dominated the first few decades of cognitive psychology. In the 1980s and 1990s, connectionists held sway, but now Bayesians are on the rise.

For the hardest problems—the ones we really want to solve but haven't been able to, like curing cancer—pure nature-inspired approaches are probably too uninformed to succeed, even given massive amounts of data. We can in principle learn a complete model of a cell's metabolic networks by a combination of structure search, with or without crossover, and parameter learning via backpropagation, but there are too many bad local optima to get stuck in. We need to reason with larger chunks, assembling and reassembling them as needed and using inverse deduction to fill in the gaps. And we need our learning to be guided by the goal of optimally diagnosing cancer and finding the best drugs to cure it.

Optimal learning is the Bayesians' central goal, and they are in no doubt that they've figured out how to reach it. This way, please . . .

CHAPTER SIX

In the Church of the Reverend Bayes

The dark hulk of the cathedral rises from the night. Light pours from its stained-glass windows, projecting intricate equations onto the streets and buildings beyond. As you approach, you can hear chanting inside. It seems to be Latin, or perhaps math, but the Babel fish in your ear translates it into English: "Turn the crank! Turn the crank!" Just as you enter, the chant dissolves into an "Aaaah!" of satisfaction, and a murmur of "The posterior! The posterior!" You peek through the crowd. A massive stone tablet towers above the altar with a formula engraved on it in ten-foot letters:

$$P(A|B) = P(A) \, P(B|A) \, / \, P(B)$$

As you stare uncomprehendingly at it, your Google Glass helpfully flashes: "Bayes' theorem." Now the crowd starts to chant "More data! More data!" A stream of sacrificial victims is being inexorably pushed toward the altar. Suddenly, you realize that you're in the middle of it—too late. As the crank looms over you, you scream, "No! I don't want to be a data point! Let me gooooo!"

You wake up in a cold sweat. Lying on your lap is a book entitled *The Master Algorithm*. Shaking off the nightmare, you resume reading where you had left off.

The theorem that runs the world

The path to optimal learning begins with a formula that many people have heard of: Bayes' theorem. But here we'll see it in a whole new light and realize that it's vastly more powerful than you'd guess from its everyday uses. At heart, Bayes' theorem is just a simple rule for updating your degree of belief in a hypothesis when you receive new evidence: if the evidence is consistent with the hypothesis, the probability of the hypothesis goes up; if not, it goes down. For example, if you test positive for AIDS, your probability of having it goes up. Things get more interesting when you have many pieces of evidence, such as the results of multiple tests. To combine them all without suffering a combinatorial explosion, we need to make simplifying assumptions. Things get even more interesting when we consider many hypotheses at once, such as all the different possible diagnoses for a patient. Computing the probability of each disease from the patient's symptoms in a reasonable amount of time can take a lot of smarts. Once we know how to do all these things, we'll be ready to learn the Bayesian way. For Bayesians, learning is "just" another application of Bayes' theorem, with whole models as the hypotheses and the data as the evidence: as you see more data, some models become more likely and some less, until ideally one model stands out as the clear winner. Bayesians have invented fiendishly clever kinds of models. So let's get started.

Thomas Bayes was an eighteenth-century English clergyman who, without realizing it, became the center of a new religion. You may well ask how that could happen, until you notice that it happened to Jesus, too: Christianity as we know it was invented by Saint Paul, while Jesus saw himself as the pinnacle of the Jewish faith. Similarly, Bayesianism as we know it was invented by Pierre-Simon de Laplace, a Frenchman who was born five decades after Bayes. Bayes was the preacher who first described a new way to think about chance, but it was Laplace who codified those insights into the theorem that bears Bayes's name.

One of the greatest mathematicians of all time, Laplace is perhaps best known for his dream of Newtonian determinism:

> *An intelligence that, at a given instant, could comprehend all the forces by which nature is animated and the respective situation of the beings that make it up, if moreover it were vast enough to submit these data to analysis, would encompass in the same formula the movements of the greatest bodies of the universe and those of the lightest atoms. For such an intelligence nothing would be uncertain, and the future, like the past, would be open to its eyes.*

This is ironic, since Laplace was also the father of probability theory, which he believed was just common sense reduced to calculation. At the heart of his explorations in probability was a preoccupation with Hume's question. For example, how do we know the sun will rise tomorrow? It has done so every day until today, but that's no guarantee it will continue. Laplace's answer had two parts. The first is what we now call the principle of indifference, or principle of insufficient reason. We wake up one day—at the beginning of time, let's say, which for Laplace was five thousand years or so ago—and after a beautiful afternoon, we see the sun go down. Will it come back? We've never seen the sun rise, and there is no particular reason to believe it will or won't. Therefore we should consider the two scenarios equally likely and say that the sun will rise again with a probability of one-half. But, Laplace went on, if the past is any guide to the future, every day that the sun rises should increase our confidence that it will continue to do so. After five thousand years, the probability that the sun will rise yet again tomorrow should be very close to one, but not quite there, since we can never be completely certain. From this thought experiment, Laplace derived his so-called rule of succession, which estimates the probability that the sun will rise again after having risen n times as $(n + 1) / (n + 2)$. When $n = 0$, this is just ½; and as *n* increases, so does the probability, approaching 1 when *n* approaches infinity.

This rule arises from a more general principle. Suppose you awake in the middle of the night on a strange planet. Even though all you can see is the starry sky, you have reason to believe that the sun will rise at some point, since most planets revolve around themselves and their sun. So your estimate of the corresponding probability should be greater than one-half (two-thirds, say). We call this the *prior probability* that the sun will rise, since it's prior to seeing any evidence. It's not based on counting the number of times the sun has risen on this planet in the past, because you weren't there to see it; rather, it reflects your a priori beliefs about what will happen, based on your general knowledge of the universe. But now the stars start to fade, so your confidence that the sun does rise on this planet goes up, based on your experience on Earth. Your confidence is now a *posterior probability*, since it's after seeing some evidence. The sky begins to lighten, and the posterior probability takes another leap. Finally, a sliver of the sun's bright disk appears above the horizon and perhaps catches "the Sultan's turret in a noose of light," as in the opening verse of the *Rubaiyat*. Unless you're hallucinating, it is now certain that the sun will rise.

The crucial question is exactly how the posterior probability should evolve as you see more evidence. The answer is Bayes' theorem. We can think of it in terms of cause and effect. Sunrise causes the stars to fade and the sky to lighten, but the latter is stronger evidence of daybreak, since the stars could fade in the middle of the night due to, say, fog rolling in. So the probability of sunrise should increase more after seeing the sky lighten than after seeing the stars fade. In mathematical notation, we say that $P(sunrise \mid lightening\text{-}sky)$, the conditional probability of sunrise given that the sky is lightening, is greater than $P(sunrise \mid fading\text{-}stars)$, its conditional probability given that the stars are fading. According to Bayes' theorem, the more likely the effect is given the cause, the more likely the cause is given the effect: if $P(lightening\text{-}sky \mid sunrise)$ is higher than $P(fading\text{-}stars \mid sunrise)$, perhaps because some planets are far enough from their sun that the stars still shine after sunrise, then *P(sunrise | lightening sky)* is also higher than *P(sunrise | fading-stars)*.

This is not the whole story, however. If we observe an effect that would happen even without the cause, then surely that's not much evidence of the cause being present. Bayes' theorem incorporates this by saying that *P(cause | effect)* goes down with *P(effect)*, the prior probability of the effect (i.e., its probability in the absence of any knowledge of the causes). Finally, other things being equal, the more likely a cause is a priori, the more likely it should be a posteriori. Putting all of these together, Bayes' theorem says that

$$P(cause \mid effect) = P(cause) \times P(effect \mid cause) / P(effect).$$

Replace *cause* by *A* and *effect* by *B* and omit the multiplication sign for brevity, and you get the ten-foot formula in the cathedral.

That's just a statement of the theorem, not a proof, of course. But the proof is surprisingly simple. We can illustrate it with an example from medical diagnosis, one of the "killer apps" of Bayesian inference. Suppose you're a doctor, and you've diagnosed a hundred patients in the last month. Fourteen of them had the flu, twenty had a fever, and eleven had both. The conditional probability of fever given flu is therefore eleven out of fourteen, or 11/14. Conditioning reduces the size of the universe that we're considering, in this case from all patients to only patients with the flu. In the universe of all patients, the probability of fever is 20/100; in the universe of flu-stricken patients, it's 11/14. The probability that a patient has the flu *and* a fever is the fraction of patients that have the flu times the fraction of *those* that have a fever: *P(flu, fever)* = *P(flu)* × *P(fever | flu)* = 14/100 × 11/14 = 11/100. But we could equally well have done this the other way around: *P(flu, fever)* = *P(fever)* × *P(flu | fever)*. Therefore, since they're both equal to *P(flu,fever)*, *P(fever)* × *P(flu | fever)* = *P(flu)* × *P(fever | flu)*. Divide both sides by *P(fever)*, and you get *P(flu | fever)* = *P(flu)* × *P(fever | flu)* / *P(fever)*. That's it! That's Bayes' theorem, with flu as the cause and fever as the effect.

Humans, it turns out, are not very good at Bayesian inference, at least when verbal reasoning is involved. The problem is that we tend to neglect the cause's prior probability. If you test positive for HIV, and the test only gives 1 percent false positives, should you panic? At first sight, it seems like your chances of having AIDS are now 99 percent. Yikes! But let's keep a cool head and apply Bayes' theorem step-by-step: *P(HIV | positive) = P(HIV) × P(positive | HIV) / P(positive)*. *P(HIV)* is the prevalence of HIV in the general population, which is about 0.3 percent in the United States. *P(positive)* is the probability that the test comes out positive whether or not you have AIDS; let's say that's 1 percent. So *P(HIV | positive)* = 0.003 × 0.99 / 0.01 = 0.297. That's very different from 0.99! The reason is that HIV is rare in the general population. The test coming out positive increases your chances of having AIDS by two orders of magnitude, but they're still less than half. If you test positive for HIV, the right thing to do is to stay calm and take another, more definitive test. Chances are you'll be fine.

Bayes' theorem is useful because what we usually know is the probability of the effects given the causes, but what we want to know is the probability of the causes given the effects. For example, we know what percentage of flu patients have a fever, but what we really want to know is how likely a patient with a fever is to have the flu. Bayes' theorem lets us go from one to the other. Its significance extends far beyond that, however. For Bayesians, this innocent-looking formula is the *F = ma* of machine learning, the foundation from which a vast number of results and applications flow. And whatever the Master Algorithm is, it must be "just" a computational implementation of Bayes' theorem. I put *just* in quotes because implementing Bayes' theorem on a computer turns out to be fiendishly hard for all but the simplest problems, for reasons that we're about to see.

Bayes' theorem as a foundation for statistics and machine learning is bedeviled not just by computational difficulty but also by extreme controversy. You might be forgiven for wondering why: Isn't it a straightforward consequence of the notion of conditional probability, as we saw in the flu example? Indeed, no one has a problem with the formula itself. The controversy is in how Bayesians obtain the probabilities that go into it and what those probabilities mean. For most statisticians, the only legitimate way to estimate probabilities is by counting how often the corresponding events occur. For example, the probability of fever is 0.2 because twenty out of one hundred observed patients had it. This is the "frequentist" interpretation of probability, and the dominant school of thought in statistics takes its name from it. But notice that in the sunrise example, and in Laplace's principle of indifference, we did something different: we pulled a probability out of thin air. What exactly justifies assuming a priori that the probability the sun will rise is one-half, or two-thirds, or whatever? Bayesians' answer is that a probability is not a frequency but a subjective degree of belief. Therefore it's up to you what you make it, and all that Bayesian inference lets you do is update your prior beliefs with new evidence to obtain your posterior beliefs (also known as "turning the Bayesian crank"). Bayesians' devotion to this idea is near religious, enough to withstand two hundred years of attacks and counting. And with the appearance on the stage of computers powerful enough to do Bayesian inference, and the massive data sets to go with it, they're beginning to gain the upper hand.

All models are wrong, but some are useful

In reality, a doctor doesn't diagnose the flu just based on whether you have a fever; she takes a whole bunch of symptoms into account, including whether you have a cough, a sore throat, a runny nose, a headache, chills, and so on. So what we really need to compute is $P(flu \mid fever, cough, sore throat, runny nose, headache, chills, \ldots)$. By Bayes' theorem, we know that this is proportional to $P(fever, cough, sore throat, runny nose, headache, chills, \ldots \mid flu)$. But now we run into a problem. How are we supposed to estimate this probability? If each symptom is a Boolean variable (you either have it or you don't) and the doctor takes n symptoms into account, a patient could have 2^n possible combinations of symptoms. If we have, say, twenty symptoms and a database of ten thousand patients, we've only seen a small fraction of the roughly one million possible combinations. Worse still, to accurately estimate the probability of a particular combination, we need at least tens of observations of it, meaning the database would need to include tens of millions of patients. Add another ten symptoms, and we'd need more patients than there are people on Earth. With a hundred symptoms, even if we were somehow able to magically get the data, there wouldn't be enough space on all the hard disks in the world to store all the probabilities. And if a patient walks in with a combination of symptoms we haven't seen before, we won't know how to diagnose him. We're face-to-face with our old foe: the combinatorial explosion.

Therefore we do what we always have to do in life: compromise. We make simplifying assumptions that whittle the number of probabilities we have to estimate down to something manageable. A very simple and popular assumption is that all the effects are independent given the cause. This means that, for example, having a fever doesn't change how likely you are to also have a cough, if we already know you have the flu. Mathematically, this is saying that $P(fever, cough \mid flu)$ is just $P(fever \mid flu) \times P(cough \mid flu)$. Lo and behold: each of these is easy to estimate from a small number of observations. In fact, we did it for fever in the previous section, and it would be no different for cough or any other symptom. The number of observations we need no longer goes up exponentially with the number of symptoms; in fact, it doesn't go up at all.

Notice that we're only saying that fever and cough are independent given that you have the flu, not overall. Clearly, if we don't know whether you have the flu, fever and cough are highly correlated, since you're much more likely to have a cough if you already have a fever. $P(fever, cough)$ is *not* equal to $P(fever) \times P(cough)$. All we're saying is that, if we know you have the flu, knowing whether you have a fever gives us no *additional* information about whether you have a cough. Likewise, if you don't know the sun is about to rise and you see the stars fade, your expectation that the sky will lighten increases; but if you already know that sunrise is imminent, seeing the stars fade makes no difference.

Notice also that it's only thanks to Bayes' theorem that we were able to pull off this trick. If we wanted to directly estimate $P(flu \mid fever, cough, etc.)$, without first turning it into $P(fever, cough, etc. \mid flu)$ using the theorem, we'd still need an exponential number of probabilities, one for each combination of symptoms and flu/not flu.

A learner that uses Bayes' theorem and assumes the effects are independent given the cause is called a Naïve Bayes classifier. That's because, well, that's such a naïve assumption. In reality, having a fever makes having a cough more likely, even if you already know you have the flu, because (for example) it makes you more likely to have a bad flu. But machine learning is the art of making false assumptions and getting away with it. As the statistician George Box famously put it: "All models are wrong, but some are useful." An oversimplified model that you have enough data to estimate is better than a perfect one that you don't. It's astonishing how simultaneously very wrong and very useful some models can be. The economist Milton Friedman even argued in a highly influential essay that the best theories are the most oversimplified, provided their predictions are accurate, because they explain the most with the least. That seems to me like a bridge too far, but it illustrates that, counter to Einstein's dictum, science often progresses by making things as simple as possible, and then some.

No one is sure who invented the Naïve Bayes algorithm. It was mentioned without attribution in a 1973 pattern recognition textbook, but it only took off in the 1990s, when researchers noticed that, surprisingly, it was often more accurate than much more sophisticated learners. I was a graduate student at the time, and when I belatedly decided to include Naïve Bayes in my experiments, I was shocked to find it did better than all the other algorithms I was comparing, save one—luckily, the algorithm I was developing for my thesis, or I might not be here now.

Naïve Bayes is now very widely used. For example, it forms the basis of many spam filters. It all began when David Heckerman, a prominent Bayesian researcher who is also a medical doctor, had the idea of treating spam as a disease whose symptoms are the words in the e-mail: *Viagra* is a symptom, and so is *free*, but your best friend's first name probably signals a legit e-mail. We can then use Naïve Bayes to classify e-mails into spam and nonspam, provided spammers generate e-mails by picking words at random. That's a ridiculous assumption, of course: it would only be true if sentences had no syntax and no content. But that summer Mehran Sahami, then a Stanford graduate student, tried it out during an internship at Microsoft Research, and it worked great. When Bill Gates asked Heckerman how this could be, he pointed out that to identify spam you don't need to understand the details of the message; it's enough to get the gist of it by seeing which words it contains.

A basic search engine also uses an algorithm quite similar to Naïve Bayes to decide which web pages to return in answer to your query. The main difference is that, instead of spam/not-spam, it's trying to predict relevant/not-relevant. The list of prediction problems Naïve Bayes has been applied to is practically endless. Peter Norvig, director of research at Google, told me at one point that it was the most widely used learner there, and Google uses machine learning in every nook and cranny of what it does. It's not hard to see why Naïve Bayes would be popular among Googlers. Surprising accuracy aside, it scales great; learning a Naïve Bayes classifier is just a matter of counting how many times each attribute co-occurs with each class and takes barely longer than reading the data from disk.

You could even use Naïve Bayes, tongue-in-cheek, on a much larger scale than Google's: to model the whole universe. Indeed, if you believe in an omnipotent God, then you can model the universe as a vast Naïve Bayes distribution where everything that happens is independent given God's will. The catch, of course, is that we can't read God's mind, but in Chapter 8 we'll investigate how to learn Naïve Bayes models even when we don't know the classes of the examples.

It might not seem so at first, but Naïve Bayes is closely related to the perceptron algorithm. The perceptron adds weights and Naïve Bayes multiplies probabilities, but if you take a logarithm, the latter reduces to the former. Both can be seen as generalizations of simple *If . . . then . . .* rules, where each antecedent can count more or less toward the conclusion instead of being "all or none." This is just one example of the deeper connections among learners that hint at a Master Algorithm. You may not consciously know Bayes' theorem (well, now you do), but in a way every one of the ten billion neurons in your brain is a tiny instance of it.

Naïve Bayes is a good conceptual model of a learner to use when reading the press: it captures the pairwise correlation between each input and the output, which is often all that's needed to understand references to learning algorithms in news stories. But machine learning is not just pairwise correlations, of course, any more than the brain is just one neuron. The real action begins when we look for more complex patterns.

From *Eugene Onegin* to Siri

In 1913, on the eve of World War I, the Russian mathematician Andrei Markov published a paper applying probability to, of all things, poetry. In it, he modeled a classic of Russian literature, Pushkin's *Eugene Onegin,* using what we now call a Markov chain. Rather than assume that each letter was generated at random independently of the rest, he introduced a bare minimum of sequential structure: he let the probability of each letter depend on the letter immediately preceding it. He showed that, for example, vowels and consonants tend to alternate, so if you see a consonant, the next letter (ignoring punctuation and white space) is much more likely to be a vowel than it would be if letters were independent. This may not seem like much, but in the days before computers, it required spending hours manually counting characters, and Markov's idea was quite new. If $Vowel_i$ is a Boolean variable that's true if the ith letter of *Eugene Onegin* is a vowel and false if it's a consonant, we can represent Markov's model with a chain-like graph like this, with an arrow between two nodes indicating a direct dependency between the corresponding variables:

Markov assumed (wrongly but usefully) that the probabilities are the same at every position in the text. Thus we need to estimate only three probabilities: $P(Vowel_1 = True)$, $P(Vowel_{i+1} = True \mid Vowel_i = True)$, and $P(Vowel_{i+1} = True \mid Vowel_i = False)$. (Since probabilities sum to one, from these we can immediately obtain $P(Vowel_1 = False)$, etc.) As with

Naïve Bayes, we can have as many variables as we want without the number of probabilities we need to estimate going through the roof, but now the variables actually depend on each other.

If we measure not just the probability of vowels versus consonants, but the probability of each letter in the alphabet following each other, we can have fun generating new texts with the same statistics as *Onegin*: choose the first letter, then choose the second based on the first, and so on. The result is complete gibberish, of course, but if we let each letter depend on several previous letters instead of just one, it starts to sound more like the ramblings of a drunkard, locally coherent even if globally meaningless. Still not enough to pass the Turing test, but models like this are a key component of machine-translation systems, like Google Translate, which lets you see the whole web in English (or almost), regardless of the language the pages were originally written in.

PageRank, the algorithm that gave rise to Google, is itself a Markov chain. Larry Page's idea was that web pages with many incoming links are probably more important than pages with few, and links from important pages should themselves count for more. This sets up an infinite regress, but we can handle it with a Markov chain. Imagine a web surfer going from page to page by randomly following links: the states of this Markov chain are web pages instead of characters, making it a vastly larger problem, but the math is the same. A page's score is then the fraction of the time the surfer spends on it, or equivalently, his probability of landing on the page after wandering around for a long time.

Markov chains turn up everywhere and are one of the most intensively studied topics in mathematics, but they're still a very limited kind of probabilistic model. We can go one step further with a model like this:

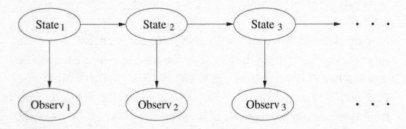

The states form a Markov chain, as before, but we don't get to see them; we have to infer them from the observations. This is called a hidden Markov model, or HMM for short. (Slightly misleading, because it's the states that are hidden, not the model.) HMMs are at the heart of speech-recognition systems like Siri. In speech recognition, the hidden states are written words, the observations are the sounds spoken to Siri, and the goal is to infer the words from the sounds. The model has two components: the probability of the next word given the current one, as in a Markov chain, and the probability of hearing various sounds given the word being pronounced. (How exactly to do the inference is a fascinating problem that we'll turn to after the next section.)

Siri aside, you use an HMM every time you talk on your cell phone. That's because your words get sent over the air as a stream of bits, and the bits get corrupted in transit. The HMM then figures out the intended bits (hidden state) from the ones received (observations), which it should be able to do as long as not too many bits got mangled.

HMMs are also a favorite tool of computational biologists. A protein is a sequence of amino acids, and DNA is a sequence of bases. If we want to predict, for example, how a protein will fold into a 3-D shape, we can treat the amino acids as the observations and the type of fold at each point as the hidden state. Similarly, we can use an HMM to identify the sites in DNA where gene transcription is initiated and many other properties.

If the states and observations are continuous variables instead of discrete ones, the HMM becomes what's known as a Kalman filter. Economists use Kalman filters to remove noise from time series of quantities like GDP, inflation, and unemployment. The "true" GDP values are the hidden states; at each time step, the true value should be similar to the observed one, but also to the previous true value, since the economy seldom makes abrupt jumps. The Kalman filter trades off these two, yielding a smoother curve that still accords with the observations. When a missile cruises to its target, it's a Kalman filter that keeps it on track. Without it, there would have been no man on the moon.

Everything is connected, but not directly

HMMs are good for modeling sequences of all kinds, but they're still a far cry from the flexibility of the symbolists' *If . . . then . . .* rules, where anything can appear as an antecedent, and a rule's consequent can in turn be an antecedent in any downstream rule. If we allow such an arbitrary structure in practice, however, the number of probabilities we need to learn blows up. For a long time no one knew how to square this circle, and researchers resorted to ad-hoc schemes, like attaching confidence estimates to rules and somehow combining them. If A implies B with confidence 0.8 and B implies C with confidence 0.7, then perhaps A implies C with confidence 0.8 × 0.7.

The problem with these schemes is that they can go badly awry. From the two perfectly reasonable rules *If the sprinkler is on, then the grass is wet* and *If the grass is wet, then it rained*, I can infer the nonsensical rule *If the sprinkler is on, then it rained*. A more insidious problem is that with confidence-rated rules we're prone to double-counting evidence. Suppose you read in the *New York Times* that aliens have landed. Maybe it's a prank, even though it's not April 1. But now you see the same headline in the *Wall Street Journal*, *USA Today,* and the *Washington Post*. You start to panic, like the listeners to Orson Welles's infamous *War of the Worlds* radio broadcast who didn't realize it was a dramatization. If, however, you check the fine print and notice that all four newspapers got the story from the Associated Press, you go back to suspecting it's a prank, this time by an AP reporter. Rule systems have no way of dealing with this, and neither does Naïve Bayes. If it uses features like *Reported in the* New York Times as predictors that a news story is true, all it can do is add *Reported by AP,* which only makes things worse.

The breakthrough came in the early 1980s, when Judea Pearl, a professor of computer science at the University of California, Los Angeles, invented a new representation: Bayesian networks. Pearl is one of the most distinguished computer scientists in the world, his methods having swept through machine learning, AI, and many other fields. He won the Turing Award, the Nobel Prize of computer science, in 2012.

Pearl realized that it's OK to have a complex network of dependencies among random variables, provided each variable depends directly on only a few others. We can represent these dependencies with a graph like the ones we saw for Markov chains and HMMs, except now the graph can have any structure (as long as the arrows don't form closed loops). One of Pearl's favorite examples is burglar alarms. The alarm at your house should go off if a burglar attempts to break in, but it could also be triggered by an earthquake. (In Los Angeles, where Pearl lives, earthquakes are almost as frequent as burglaries.) If you're working late one night and your neighbor Bob calls to say he just heard your alarm go off, but your neighbor Claire doesn't, should you call the police? Here's the graph of dependencies:

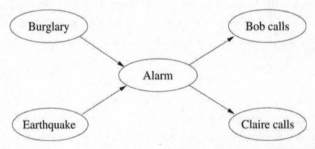

If there's an arrow from one node to another in the graph, we say that the first node is a *parent* of the second. So *Alarm*'s parents are *Burglary* and *Earthquake*, and *Alarm* is the sole parent of *Bob calls* and *Claire calls*. A Bayesian network is a graph of dependencies like this, together with a table for each variable, giving its probability for each combination of values of its parents. For *Burglary* and *Earthquake* we only need one probability each, since they have no parents. For *Alarm* we need four: the probability that it goes off even if there's no burglary or earthquake, the probability that it goes off if there's a burglary and no earthquake, and so on. For *Bob calls* we need two probabilities (given alarm and given no alarm), and similarly for Claire.

Here's the crucial point: Bob calling depends on *Burglary* and *Earthquake*, but only through *Alarm*. Bob's call is *conditionally independent* of *Burglary* and *Earthquake* given *Alarm*, and so is Claire's. If the alarm doesn't go off, your neighbors sleep soundly, and the burglar proceeds undisturbed. Also, Bob and Claire are independent given *Alarm*. Without this independence structure, you'd need to learn $2^5 = 32$ probabilities, one for each possible state of the five variables. (Or 31, if you're a stickler for details, since the last one can be left implicit.) With the conditional independencies, all you need is $1 + 1 + 4 + 2 + 2 = 10$, a savings of 68 percent. And that's just in this tiny example; with hundreds or thousands of variables, the savings would be very close to 100 percent.

The first law of ecology, according to biologist Barry Commoner, is that everything is connected to everything else. That may be true, but it would also make the world impossible to understand, if not for the saving grace of conditional independence: everything is connected, but only indirectly. In order to affect me, something that happens a mile away must first affect something in my neighborhood, even if only through the propagation of light. As one wag put it, space is the reason everything doesn't happen to you. Put another way, the structure of space is an instance of conditional independence.

In the burglary example, the full table of thirty-two probabilities is never represented explicitly, but it's implicit in the collection of smaller tables and graph structure. To obtain *P(Burglary, Earthquake, Alarm, Bob calls, Claire calls)*, all I have to do is multiply *P(Burglary)*, *P(Earthquake)*, *P(Alarm | Burglary, Earthquake)*, *P(Bob calls | Alarm)*, and *P(Claire calls | Alarm)*. It's the same in any Bayesian network: to obtain the probability of a complete state, just multiply the probabilities from the corresponding lines in the individual variables' tables. So, provided the conditional independencies hold, no information is lost by switching to the more compact representation. And in this way we can easily compute the probabilities of extremely unusual states, including states that were never observed before. Bayesian networks give the lie to the common misconception that machine learning can't predict very rare events, or "black swans," as Nassim Taleb calls them.

In retrospect, we can see that Naïve Bayes, Markov chains, and HMMs are all special cases of Bayesian networks. The structure of Naïve Bayes is:

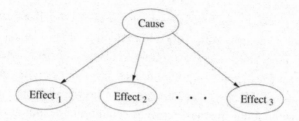

Markov chains encode the assumption that the future is conditionally independent of the past given the present. HMMs assume in addition that each observation depends only on the corresponding state. Bayesian networks are for Bayesians what logic is for symbolists: a lingua franca that allows us to elegantly encode a dizzying variety of situations and devise algorithms that work uniformly in all of them.

We can think of a Bayesian network as a "generative model," a recipe for probabilistically generating a state of the world: first decide independently whether there's a burglary and/or an earthquake, then based on that decide whether the alarm goes off, and then based on that whether Bob and Claire call. A Bayesian network tells a story: A happened, and it led to B; at the same time, C also happened, and B and C together caused D. To compute the probability of a particular story, we just multiply the probabilities of all of its different strands.

One of the most exciting applications of Bayesian networks is modeling how genes regulate each other in living cells. Billions of dollars have been spent trying to discover pairwise correlations between individual genes and specific diseases, but the yield has been disappointingly low. In retrospect, this is not so surprising: a cell's behavior is the result of complex interactions among genes and the environment, and a single gene has limited predictive power. But with Bayesian networks, we can uncover these interactions, provided we have the requisite data, and with the spread of DNA microarrays, we increasingly do.

After pioneering the application of machine learning to spam filtering, David Heckerman turned to using Bayesian networks in the fight against AIDS. The AIDS virus is a tough adversary because it mutates rapidly, making it difficult for any one vaccine or drug to pin it down for long. Heckerman noticed that this is the same cat-and-mouse game that spam filters play with spam and decided to apply a lesson he had learned there: attack the weakest link. In the case of spam, weak links include the URLs you have to use to take payment from the customer. In the case of HIV, they're small regions of the virus protein that can't change without hurting the virus. If he could train the immune system to recognize these regions and attack the cells displaying them, he just might have an AIDS vaccine. Heckerman and coworkers used a Bayesian network to help identify the vulnerable regions and developed a vaccine delivery mechanism that could teach the immune system to attack just those regions. The delivery mechanism worked in mice, and clinical trials are now in preparation.

It often happens that, even after we take all conditional independences into account, some nodes in a Bayesian network still have too many parents. Some networks are so dense with arrows that when we print them, the page turns solid black. (The physicist Mark Newman calls them "ridiculograms.") A doctor needs to simultaneously diagnose all the possible diseases a patient could have, not just one, and every disease is a parent of many different symptoms. A fever could be caused by any number of conditions besides the flu, but it's hopeless to try to predict its probability given every possible combination of conditions. All is not lost. Instead of a table specifying the node's conditional probability for every state of its parents, we can learn a simpler distribution. The most popular choice is a probabilistic version of the logical OR operation: any cause alone can provoke a fever, but each cause has a certain probability of failing to do so, even if it's usually sufficient. Heckerman and others have learned Bayesian networks that diagnose hundreds of infectious diseases in this way. Google uses a giant Bayesian network of this type in its AdSense system for automatically choosing ads to place on web pages. The network relates a million content variables to each other and to twelve million words and phrases via over three hundred million arrows, all learned from a hundred billion text snippets and search queries.

On a lighter note, Microsoft's Xbox Live uses a Bayesian network to rate players and match players of similar skill. The outcome of a game is a probabilistic function of the opponents' skill levels, and using Bayes' theorem we can infer a player's skill from the outcomes of his games.

The inference problem

There's a big snag in all of this, unfortunately. Just because a Bayesian network lets us compactly represent a probability distribution doesn't mean we can also reason efficiently with it. Suppose you want to compute *P(Burglary | Bob called, Claire didn't)*. By Bayes' theorem, you know this is just *P(Burglary) P(Bob called, Claire didn't | Burglary) / P(Bob called, Claire didn't)*, or equivalently, *P(Burglary, Bob called, Claire didn't) / P(Bob called, Claire didn't)*. If you had the full table with the probabilities of all states, you could obtain both of these probabilities by adding up the corresponding lines in the table. For example, *P(Bob called, Claire didn't)* is the sum of the probabilities of all the lines where Bob calls and Claire doesn't. But the Bayesian network doesn't give you the full table. You could always construct it from the individual tables, but that takes exponential time and space. What we really want is to compute *P(Burglary | Bob called, Claire didn't)* without building the full table. That, in a nutshell, is the problem of inference in Bayesian networks.

In many cases we can do this and avoid the exponential blowup. Suppose you're leading a platoon in single file through enemy territory in the dead of night, and you want to make sure that all your soldiers are still with you. You could stop and count them yourself, but that wastes too much time. A cleverer solution is to just ask the first soldier behind you: "How many soldiers are behind you?" Each soldier asks the next the same question, until the last one says "None." The next-to-last soldier can now say "One," and so on all the way back to the first soldier, with each soldier adding one to the number of soldiers behind him. Now you know how many soldiers are still with you, and you didn't even have to stop.

Siri uses the same idea to compute the probability that you just said, "Call the police" from the sounds it picked up from the microphone. Think of "Call the police" as a platoon of words marching across the page in single file. *Police* wants to know its probability, but for that it needs to know the probability of *the*; and *the* in turn needs to know the probability of *call*. So *call* computes its probability and passes it on to *the*, which does the same and passes the result to *police*. Now *police* knows its probability, duly influenced by every word in the sentence, but we never had to construct the full table of eight possibilities (the first word is *call* or isn't, the second is *the* or isn't, and the third is *police* or isn't). In reality, Siri considers all words that could appear in each position, not just whether the first word is *call* or not and so on, but the algorithm is the same. Perhaps Siri thinks, based on the sounds, that the first word was either *call* or *tell*, the second was *the* or *her*, and the third was *police* or *please*. Individually, perhaps the most likely words are *call*, *the*, and *please*. But that forms the nonsensical sentence "Call the please," so taking the other words into account, Siri concludes that the sentence is really "Call the police." It makes the call, and with luck the police get to your house in time to catch the burglar.

The same idea still works if the graph is a tree instead of a chain. If instead of a platoon you're in command of a whole army, you can ask each of your company commanders how many soldiers are behind him and add up their answers. Each company commander in turn asks each of his platoon commanders, and so on. But if the graph forms loops, you're in trouble. If there's a liaison officer who's a member of two platoons, he gets counted twice; in fact, everyone behind him gets counted twice. This is what happens in the "aliens have landed" scenario, if you want to compute, say, the probability of panic:

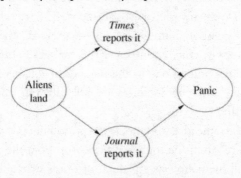

One solution is to combine *The* Times *reports it* and *The* Journal *reports it* into a single megavariable with four values: *YesYes* if they both do, *YesNo* if the *Times* reports a landing and the *Journal* doesn't, and so on. This turns the graph into a chain of three variables, and all is well. However, every time you add a news source, the number of values of the megavariable doubles. If instead of two news sources you have fifty, the megavariable has 2^{50} values. So this method can only get you so far, and no other known method does any better.

The problem is worse than it seems, because Bayesian networks in effect have "invisible" arrows to go along with the visible ones. *Burglary* and *Earthquake* are a priori independent, but the alarm going off entangles them: the alarm makes you suspect a burglary, but if now you hear on the radio that there's been an earthquake, you assume that's what caused the alarm. The earthquake has *explained away* the alarm, making a burglary less likely, and the two are therefore dependent. In a Bayesian network, all parents of the same variable are interdependent in this way, and this in turn introduces further dependencies, making the resulting graph often much denser than the original one.

The crucial question for inference is whether you can make the filled-in graph "look like a tree" without the trunk getting too thick. If the megavariable in the trunk has too many possible values, the tree grows out of control until it covers the whole planet, like the baobabs in *The Little Prince*. In the tree of life, each species is a branch, but inside each branch is a graph, with each creature having two parents, four grandparents, some number of offspring, and so on. The "thickness" of a branch is the size of the species' population. When the branches are too thick, our only choice is to resort to approximate inference.

One solution, left as an exercise by Pearl in his book on Bayesian networks, is to pretend the graph has no loops and just keep propagating probabilities back and forth until they converge. This is known as loopy belief propagation, both because it works on graphs with loops and because it's a crazy idea. Surprisingly, it turns out to work quite well in many cases. For instance, it's a state-of-the art method for wireless communication, with the random variables being the bits in the message, encoded in a clever way. But loopy belief propagation can also converge to the wrong answers or oscillate forever. Another solution, which originated in physics but was imported into machine learning and greatly extended by Michael Jordan and others, is to approximate an intractable distribution with a tractable one and optimize the latter's parameters to make it as close as possible to the former.

The most popular option, however, is to drown our sorrows in alcohol, get punch drunk, and stumble around all night. The technical term for this is *Markov chain Monte Carlo*, or MCMC for short. The "Monte Carlo" part is because the method involves chance, like a visit to the eponymous casino, and the "Markov chain" part is because it involves taking a sequence of steps, each of which depends only on the previous one. The idea in MCMC is to do a random walk, like the proverbial drunkard, jumping from state to state of the network in such a way that, in the long run, the number of times each state is visited is proportional to its probability. We can then estimate the probability of a burglary, say, as the fraction of times we visited a state where there was a burglary. A "well-behaved" Markov chain converges to a stable distribution, so after a while it always gives approximately the same answers. For example, when you shuffle a deck of cards, after a while all card orders are equally likely, no matter the initial order; so you know that if there are n possible orders, the probability of each one is $1/n$. The trick in MCMC is to design a Markov chain that converges to the distribution of our Bayesian network. One easy option is to repeatedly cycle through the variables, sampling each one according to its conditional probability given the state of its neighbors. People often talk about MCMC as a kind of simulation, but it's not: the Markov chain does not simulate any real process; rather, we concocted it to efficiently generate samples from a Bayesian network, which is itself not a sequential model.

The origins of MCMC go all the way back to the Manhattan Project, when physicists needed to estimate the probability that neutrons would collide with atoms and set off a chain reaction. But in more recent decades, it has sparked such a revolution that it's often considered one of the most important algorithms of all time. MCMC is good not just for computing probabilities but for integrating any function. Without it, scientists were limited to functions they could integrate analytically, or to well-behaved, low-dimensional integrals they could approximate as a series of trapezoids. With MCMC, they're free to build complex models, knowing the computer will do the heavy lifting. Bayesians, for one, probably have MCMC to thank for the rising popularity of their methods more than anything else.

On the downside, MCMC is often excruciatingly slow to converge, or fools you by looking like it's converged when it hasn't. Real probability distributions are usually very peaked, with vast wastelands of minuscule probability punctuated by sudden Everests. The Markov chain then converges to the nearest peak and stays there, leading to very biased probability estimates. It's as if the drunkard followed the scent of alcohol to the nearest tavern and stayed there all night, instead of wandering all around the city like we wanted him to. On the other hand, if instead of using a Markov chain we just generated independent samples, like simpler Monte Carlo methods do, we'd have no scent to follow and probably wouldn't even find that first tavern; it would be like throwing darts at a map of the city, hoping they land smack dab on the pubs.

Inference in Bayesian networks is not limited to computing probabilities. It also includes finding the most probable explanation for the evidence, such as the disease that best explains the symptoms or the words that best explain the sounds Siri heard. This is not the same as just picking the most probable word at each step, because words that are individually likely given their sounds may be unlikely to occur together, as in the "Call the please" example. However, similar kinds of algorithms also work for this task (and they are, in fact, what most speech recognizers use). Most importantly, inference includes making the best decisions, guided not just by the probabilities of different outcomes but also by the corresponding costs (or utilities, to use the technical term). The cost of ignoring an e-mail from your boss asking you to do something by tomorrow is much greater than the cost of seeing a piece of spam, so often it's better to let an e-mail through even if it does seem fairly likely to be spam.

Driverless cars and other robots are a prime example of probabilistic inference in action. As the car drives around, it simultaneously builds up a map of the territory and figures out its location on it with increasing certainty. According to a recent study, London taxi drivers grow a larger posterior hippocampus, a brain region involved in memory and map making, as they learn the layout of the city. Perhaps they use similar probabilistic inference algorithms, with the notable difference that in the case of humans, drinking doesn't seem to help.

Learning the Bayesian way

Now that we know how to (more or less) solve the inference problem, we're ready to learn Bayesian networks from data, because for Bayesians learning is just another kind of probabilistic inference. All you have to do is apply Bayes' theorem with the hypotheses as the possible causes and the data as the observed effect:

$$P(hypothesis \mid data) = P(hypothesis) \times P(data \mid hypothesis) / P(data)$$

The hypothesis can be as complex as a whole Bayesian network, or as simple as the probability that a coin will come up heads. In the latter case, the data is just the outcome of a series of coin flips. If, say, we obtain seventy heads in a hundred flips, a frequentist would estimate the probability of heads as 0.7. This is justified by the so-called maximum likelihood principle: of all the possible probabilities of heads, 0.7 is the one under which seeing seventy heads in a hundred flips is most likely. The likelihood of a hypothesis is $P(data \mid hypothesis)$, and the principle says we should pick the hypothesis that maximizes it. Bayesians do something more subtle, though. They point out that we never know for sure which hypothesis is the true one, and so we shouldn't just pick one hypothesis, like a value of 0.7 for the probability of heads; rather, we should compute the posterior probability of every possible hypothesis and entertain all of them when making predictions. The sum of the probabilities of all the hypotheses must be one, so if one becomes more likely, the others become less. For a Bayesian, in fact, there is no such thing as the truth; you have a prior distribution over hypotheses, after seeing the data it becomes the posterior distribution, as given by Bayes' theorem, and that's all.

This is a radical departure from the way science is usually done. It's like saying, "Actually, neither Copernicus nor Ptolemy was right; let's just predict the planets' future trajectories assuming Earth goes round the sun and vice versa and average the results."

Of course, it's a weighted average, the weight of a hypothesis being its posterior probability, so a hypothesis that explains the data better will count for more. Still, as the joke goes, being Bayesian means never having to say you're certain.

Needless to say, carrying around a multitude of hypotheses instead of just one is a huge pain. In the case of learning a Bayesian network, we're supposed to make predictions by averaging over all possible Bayesian networks, including all possible graph structures and all possible parameter values for each structure. In some cases, we can compute the average over parameters in closed form, but with varying structures we're out of luck. We have to resort to, for example, doing MCMC over the space of networks, jumping from one possible network to another as the Markov chain progresses. Combine all this complexity and computational cost with Bayesians' controversial notion that there's really no such thing as objective reality, and it's not hard to see why frequentism has dominated science for the last century.

There's a saving grace, however, and some major reasons to prefer the Bayesian way. The saving grace is that, most of the time, almost all hypotheses wind up with a tiny posterior probability, and we can safely ignore them. In fact, just considering the single most probable hypothesis is usually a very good approximation. Suppose our prior distribution for the coin flip problem is that all probabilities of heads are equally likely. The effect of seeing the outcomes of successive flips is to concentrate the distribution more and more on the hypotheses that best agree with the data. For example, if *h* ranges over the possible probabilities of heads and a coin comes out heads 70 percent of the time, we'll see something like this:

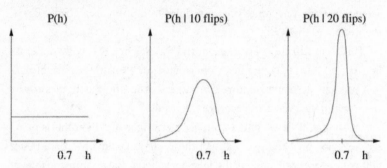

P(h)   P(h | 10 flips)   P(h | 20 flips)

0.7   h   0.7   h   0.7   h

The posterior after each flip becomes the prior for the next flip, and flip by flip, we become increasingly certain that $h = 0.7$. If we just take the single most probable hypothesis ($h = 0.7$ in this case), the Bayesian approach becomes quite similar to the frequentist one, but with one crucial difference: Bayesians take the prior *P(hypothesis)* into account, not just the likelihood *P(data | hypothesis)*. (The data prior *P(data)* can be ignored because it's the same for all hypotheses and therefore doesn't affect the choice of winner.) If we're willing to assume that all hypotheses are equally likely a priori, the Bayesian approach now reduces to the maximum likelihood principle. So Bayesians can say to frequentists: "See, what you do is a special case of what we do, but at least we make our assumptions explicit." And if the hypotheses are not equally likely a priori, maximum likelihood's implicit assumption that they are leads to the wrong answers.

This might seem like a theoretical discussion, but it has tremendous practical consequences. If we've seen only one coin flip and it came out heads, maximum likelihood says that the probability of heads must be one. This could be wildly inaccurate and leaves us woefully unprepared for the coin coming up tails. Once we've seen a lot of flips, the estimate becomes more reliable, but in many problems, we never see enough flips, no matter how big the data. Suppose the word *supercalifragilisticexpialidocious* never appears in a spam e-mail in our training data and appears once in an e-mail talking about *Mary Poppins*. A Naïve Bayes spam filter with maximum likelihood probability estimates will then decide that an e-mail containing it cannot be spam, regardless of whether every other word in the e-mail screams "Spam! Spam!" In contrast, a Bayesian would give the word a low but nonzero probability of appearing in spam, allowing the other words to override it.

The problem only gets worse if we try to learn the structure of a Bayesian network as well as its parameters. We can do this by hill climbing, starting with an empty network (no arrows), adding the arrow that most increases likelihood, and so on until no arrow causes an improvement. Unfortunately, this quickly leads to massive overfitting, with a network that assigns zero probability to all states not appearing in the data. Bayesians can do something much more interesting. They can use the prior distribution to encode experts' knowledge about the problem—their answer to Hume's question. For example, we can design an initial Bayesian network for medical diagnosis by interviewing doctors, asking them which symptoms they think depend on which diseases, and adding the corresponding arrows. This is the "prior network," and the prior distribution can penalize alternative networks by the number of arrows that they add or remove from it. But doctors are fallible, so we'll let the data override them: if the increase in likelihood from adding an arrow outweighs the penalty, we do it.

Of course, frequentists are aware of this issue, and their answer is to, for example, multiply the likelihood by a factor that penalizes more complex networks. But at this point frequentism and Bayesianism have become indistinguishable, and whether you call the scoring function "penalized likelihood" or "posterior probability" is really just a matter of taste.

Despite the convergence of frequentist and Bayesian thinking on some issues, there remains the philosophical difference about the meaning of probability. Viewing it as subjective makes many scientists queasy, but it also enables many otherwise-forbidden uses. If you're a frequentist, you can only estimate probabilities of events that can occur more than once. So a question like "What is the probability that Hillary

Clinton will beat Jeb Bush in the next presidential election?" is unanswerable, because there's never been an election pitting them against each other. But for a Bayesian, a probability is a subjective degree of belief, so he's free to make an educated guess, and the inference calculus keeps all his guesses consistent.

The Bayesian method is not just applicable to learning Bayesian networks and their special cases. (Conversely, despite their name, Bayesian networks aren't necessarily Bayesian: frequentists can learn them, too, as we just saw.) We can put a prior distribution on any class of hypotheses—sets of rules, neural networks, programs—and then update it with the hypotheses' likelihood given the data. Bayesians' view is that it's up to you what representation you choose, but then you have to learn it using Bayes' theorem. In the 1990s, they mounted a spectacular takeover of the Conference on Neural Information Processing Systems (NIPS for short), the main venue for connectionist research. The ringleaders (so to speak) were David MacKay, Radford Neal, and Michael Jordan. MacKay, a Brit who was a student of John Hopfield's at Caltech and later became chief scientific advisor to the UK's Department of Energy, showed how to learn multilayer perceptrons the Bayesian way. Neal introduced the connectionists to MCMC, and Jordan introduced them to variational inference. Finally, they pointed out that in the limit you could "integrate out" the neurons in a multilayer perceptron, leaving a type of Bayesian model that made no reference to them. Before long, the word *neural* in the title of a paper submitted to NIPS became a good predictor of rejection. Some researchers joked that the conference should change its name to BIPS, for Bayesian Information Processing Systems.

Markov weighs the evidence

But something funny happened on the way to world domination. Researchers using Bayesian models kept noticing that you got better results by tweaking the probabilities in illegal ways. For example, raising *P(words)* to some power in speech recognizers improved accuracy, but then it wasn't Bayes' theorem any more. What was going on? The culprit, it turns out, was the false independence assumptions that generative models make. The simplified graph structure makes the models learnable and is worth keeping, but then we're better off just learning the best parameters we can for the task at hand, irrespective of whether they're probabilities. The real strength of, say, Naïve Bayes is that it provides a small, informative set of features from which to predict the class and a fast, robust way to learn the corresponding parameters. In a spam filter, each feature is the occurrence of a particular word in spam, and the corresponding parameter is how often it occurs; and similarly for nonspam. Viewed in this way, Naïve Bayes can be optimal, in the sense of making the best predictions possible, even in many cases where its independence assumptions are wildly violated. When I realized this and published a paper about it in 1996, people's suspicion of Naïve Bayes melted away, helping it to take off. But it was also a step on the way to a different kind of model, which in the last two decades has increasingly replaced Bayesian networks in machine learning: Markov networks.

A Markov network is a set of features and corresponding weights, which together define a probability distribution. A feature can be as simple as *This is a ballad* or as elaborate as *This is a ballad by a hip-hop artist, with a saxophone riff and a descending chord progression*. Pandora uses a large set of features, which it calls the Music Genome Project, to select songs to play for you. Suppose we plug them into a Markov network. If you like ballads, the weight of the corresponding feature goes up, and you're more likely to hear ballads when you turn on Pandora. If you also like songs by hip-hop artists, that feature's weight also goes up. The songs you're most likely to hear are now ones that have both features, namely ballads by hip-hop artists. If you don't like ballads or hip-hop artists per se, but only enjoy them in combination, the more elaborate feature *Ballad by a hip-hop artist* is what you need. Pandora's features are handcrafted, but in Markov networks we can also learn features using hill climbing, similar to rule induction. Either way, gradient descent is a good way to learn the weights.

Like Bayesian networks, Markov networks can be represented by graphs, but they have undirected arcs instead of arrows. Two variables are connected, meaning they depend directly on each other, if they appear together in some feature, like *Ballad* and *By a hip-hop artist* in *Ballad by a hip-hop artist*.

Markov networks are a staple in many areas, such as computer vision. For instance, a driverless car needs to segment each image it sees into road, sky, and countryside. One option is to label each pixel as one of the three according to its color, but this is not nearly good enough. Images are very noisy and variable, and the car will hallucinate rocks strewn all over the roadway and patches of road in the sky. We know, however, that nearby pixels in an image are usually part of the same object, and we can introduce a corresponding set of features: for each pair of neighboring pixels, the feature is true if they belong to the same object, and false otherwise. Now images with large, contiguous blocks of road and sky are much more likely than images without, and the car goes straight instead of continually swerving left and right to avoid imaginary rocks.

Markov networks can be trained to maximize either the likelihood of the whole data or the conditional likelihood of what we want to predict given what we know. For Siri, the likelihood of the whole data is $P(words, sounds)$, and the conditional likelihood we're interested in is $P(words \mid sounds)$. By optimizing the latter, we can ignore $P(sounds)$, which is only a distraction from our goal. And since we ignore it, it can be arbitrarily complex. This is much better than HMMs' unrealistic assumption that sounds depend solely on the corresponding words, without any influence from the surroundings. In fact, if all Siri cares about is figuring out which words you just spoke, perhaps it doesn't even need to worry about probabilities; it just needs to make sure the correct words score higher than incorrect ones when it tots up the weights of their features—ideally a lot higher, just to be safe.

Analogizers took this line of reasoning to its logical conclusion, as we'll see in the next chapter. In the first decade of the new millennium, they in turn took over NIPS. Now the connectionists dominate once more, under the banner of deep learning. Some say that research goes in cycles, but it's more like a spiral, with loops winding around the direction of progress. In machine learning, the spiral converges to the Master Algorithm.

Logic and probability: The star-crossed couple

You'd think that Bayesians and symbolists would get along great, given that they both believe in a first-principles approach to learning, rather than a nature-inspired one. Far from it. Symbolists don't like probabilities and tell jokes like "How many Bayesians does it take to change a lightbulb? They're not sure. Come to think of it, they're not sure the lightbulb is burned out." More seriously, symbolists point to the high price we pay for probability. Inference suddenly becomes a lot more expensive, all those numbers are hard to understand, we have to deal with priors, and hordes of zombie hypotheses chase us around forever. The ability to compose pieces of knowledge on the fly, so dear to symbolists, is gone. Worst of all, we don't know how to put probability distributions on many of the things we need to learn. A Bayesian network is a distribution over a vector of variables, but what about distributions over networks, databases, knowledge bases, languages, plans, and computer programs, to name a few? All of these are easily handled in logic, and an algorithm that can't learn them is clearly not the Master Algorithm.

Bayesians, in turn, point to the brittleness of logic. If I have a rule like *Birds fly*, a world with even one flightless bird is impossible. If I try to patch things by adding exceptions, such as *Birds fly, unless they're penguins*, I'll never be done. (What about ostriches? Birds in cages? Dead birds? Birds with broken wings? Soaked wings?) A doctor diagnoses you with cancer, and you decide to get a second opinion. If the second doctor disagrees, you're stuck. You can't weigh the two opinions; you just have to believe them both. And then a catastrophe happens: pigs fly, perpetual motion is possible, and Earth doesn't exist—because in logic everything can be inferred from a contradiction. Furthermore, if knowledge is learned from data, I can never be sure it's true. Why do symbolists pretend otherwise? Surely Hume would frown on such insouciance.

Bayesians and symbolists agree that prior assumptions are inevitable, but they differ in the kinds of prior knowledge they allow. For Bayesians, knowledge goes in the prior distribution over the structure and parameters of the model. In principle, the parameter prior could be anything we please, but ironically, Bayesians tend to choose uninformative priors (like assigning the same probability to all hypotheses) because they're easier to compute with. In any case, humans are not very good at estimating probabilities. For structure, Bayesian networks provide an intuitive way to incorporate knowledge: draw an arrow from A to B if you think that A directly causes B. But symbolists are much more flexible: you can provide as prior knowledge to your learner anything you can encode in logic, and practically anything can be encoded in logic—provided it's black and white.

Clearly, we need both logic and probability. Curing cancer is a good example. A Bayesian network can model a single aspect of how cells function, like gene regulation or protein folding, but only logic can put all the pieces together into a coherent picture. On the other hand, logic can't deal with incomplete or noisy information, which is pervasive in experimental biology, but Bayesian networks can handle it with aplomb.

Bayesian learning works on a single table of data, where each column represents a variable (for example, the expression level of one gene) and each row represents an instance (for example, a single microarray experiment, with each gene's observed expression level). It's OK if the table has "holes" and measurement errors because we can use probabilistic inference to fill in the holes and average over the errors. But if we have more than one table, Bayesian learning is stuck. It doesn't know how to, for example, combine gene expression data with data about which DNA segments get translated into proteins, and how in turn the three-dimensional shapes of those proteins cause them to lock on to different parts of the DNA molecule, affecting the expression of other genes. In logic, we can easily write rules relating all of these aspects, and learn them from the relevant combinations of tables—but only provided the tables have no holes or errors.

Combining connectionism and evolutionism was fairly easy: just evolve the network structure and learn the parameters by backpropagation. But unifying logic and probability is a much harder problem. Attempts to do it go all the way back to Leibniz, who was a pioneer of both. Some of the best philosophers and mathematicians of the nineteenth and twentieth centuries, like George Boole and Rudolf Carnap, worked hard on it but ultimately didn't get very far. More recently, computer scientists and AI researchers have joined the fray. But as the millennium turned around, the best we had were partial successes, like adding some logical constructs to Bayesian networks. Most experts believed that unifying logic and probability was impossible. The prospects for a Master Algorithm did not look good, particularly since the existing evolutionary and connectionist algorithms couldn't deal with incomplete information or multiple data sets, either.

Luckily, we have since cracked the problem, and the Master Algorithm now looks that much closer. We'll see how we did it in Chapter 9 and take it from there. But first we need to gather a very important, still-missing piece of the puzzle: how to learn from very little data. That might seem unnecessary in these days of data deluge, but the truth is that we often find ourselves with reams of data about some parts of the problem we want to solve and almost none about others. This is where one of the most important ideas in machine learning comes in: analogy. All of the tribes we've met so far have one thing in common: they learn an explicit model of the phenomenon under consideration, whether it's a set of rules, a multilayer perceptron, a genetic program, or a Bayesian network. When they don't have enough data to do that, they're stumped. But analogizers can learn from as little as one example because they never form a model. Let's see what they do instead.

CHAPTER SEVEN

You Are What You Resemble

Frank Abagnale Jr. is one of the most notorious con men in history. Abagnale, portrayed by Leonardo DiCaprio in Spielberg's movie *Catch Me If You Can*, forged millions of dollars' worth of checks, impersonated an attorney and a college instructor, and traveled the world as a fake Pan Am pilot—all before his twenty-first birthday. But perhaps his most jaw-dropping exploit was to successfully pose as a doctor for nearly a year in late-1960s Atlanta. Practicing medicine supposedly requires many years in med school, a license, a residency, and whatnot, but Abagnale managed to bypass all these niceties and never got called on it.

Imagine for a moment trying to pull off such a stunt. You sneak into an absent doctor's office, and before long a patient comes in and tells you all his symptoms. Now you have to diagnose him, except you know nothing about medicine. All you have is a cabinet full of patient files: their symptoms, diagnoses, treatments undergone, and so on. What do you do? The easiest way out is to look in the files for the patient whose symptoms most closely resemble your current one's and make the same diagnosis. If your bedside manner is as convincing as Abagnale's, that might just do the trick. The same idea applies well beyond medicine. If you're a young president faced with a world crisis, as Kennedy was when a US spy plane revealed Soviet nuclear missiles being deployed in Cuba, chances are there's no script ready to follow. Instead, you look for historical analogs of the current situation and try to learn from them. The Joint Chiefs of Staff urged an attack on Cuba, but Kennedy, having just read *The Guns of August*, a best-selling account of the outbreak of World War I, was keenly aware of how easily that could escalate into all-out war. So he opted for a naval blockade instead, perhaps saving the world from nuclear war.

Analogy was the spark that ignited many of history's greatest scientific advances. The theory of natural selection was born when Darwin, on reading Malthus's *Essay on Population*, was struck by the parallels between the struggle for survival in the economy and in nature. Bohr's model of the atom arose from seeing it as a miniature solar system, with electrons as the planets and the nucleus as the sun. Kekulé discovered the ring shape of the benzene molecule after daydreaming of a snake eating its own tail.

Analogical reasoning has a distinguished intellectual pedigree. Aristotle expressed it in his law of similarity: if two things are similar, the thought of one will tend to trigger the thought of the other. Empiricists like Locke and Hume followed suit. Truth, said Nietzsche, is a mobile army of metaphors. Kant was also a fan. William James believed that "this sense of sameness is the very keel and backbone of our thinking." Some contemporary psychologists even argue that human cognition in its entirety is a fabric of analogies. We rely on it to find our way around a new town and to understand expressions like "see the light" and "stand tall." Teenagers who insert "like" into every sentence they say would probably, like, agree that analogy is important, dude.

Given all this, it's not surprising that analogy plays a prominent role in machine learning. It got off to a slow start, though, and was initially overshadowed by neural networks. Its first algorithmic incarnation appeared in an obscure technical report written in 1951 by two Berkeley statisticians, Evelyn Fix and Joe Hodges, and was not published in a mainstream journal until decades later. But in the meantime, other papers on Fix and Hodges's algorithm started to appear and then to multiply until it was one of the most researched in all of computer science. The nearest-neighbor algorithm, as it's called, is the first stop on our tour of analogy-based learning. The second is support vector machines, an idea that took machine learning by storm around the turn of the millennium and was only recently overshadowed by deep learning. The third and last is full-blown analogical reasoning, which has been a staple of psychology and AI for several decades, and a background theme in machine learning for nearly as long.

The analogizers are the least cohesive of the five tribes. Unlike the others, which have a strong identity and common ideals, the analogizers are more of a loose collection of researchers, united only by their reliance on similarity judgments as the basis for learning. Some, like the support vector machine folks, might even object to being brought under such an umbrella. But it's raining deep models outside, and I think they would benefit greatly from making common cause. Similarity is one of the central ideas in machine learning, and the analogizers in all their guises are its keepers. Perhaps in a future decade, machine learning will be dominated by deep analogy, combining in one algorithm the efficiency of nearest-neighbor, the mathematical sophistication of support vector machines, and the power and flexibility of analogical reasoning. (There, I just gave away one of my secret research projects.)

Match me if you can

Nearest-neighbor is the simplest and fastest learning algorithm ever invented. In fact, you could even say it's the fastest algorithm of any kind that could ever be invented. It consists of doing exactly nothing, and therefore takes zero time to run. Can't beat that. If you want to learn to recognize faces and have a vast database of images labeled face/not face, just let it sit there. Don't worry, be happy. Without knowing it, those images already implicitly form a model of what a face is. Suppose you're Facebook and you want to automatically identify faces in photos people upload as a prelude to tagging them with their friends' names. It's nice to not have to do anything, given that Facebook users upload upward of three hundred million photos per day. Applying any of the learners we've seen so far to them, with the possible exception of Naïve Bayes, would take a truckload of computers. And Naïve Bayes is not smart enough to recognize faces.

Of course, there's a price to pay, and the price comes at test time. Jane User has just uploaded a new picture. Is it a face? Nearest-neighbor's answer is: find the picture most similar to it in Facebook's entire database of labeled photos—its "nearest neighbor"—and if that picture contains a face, so does this one. Simple enough, but now you have to scan through potentially billions of photos in (ideally) a fraction of a second. Like a lazy student who doesn't bother to study for the test, nearest-neighbor is caught unprepared and has to scramble. But unlike real life, where your mother taught you to never leave until tomorrow what you can do today, in machine learning procrastination can really pay off. In fact, the entire genre of learning that nearest-neighbor is part of is sometimes called "lazy learning," and in this context there's nothing pejorative about the term.

The reason lazy learners are a lot smarter than they seem is that their models, although implicit, can in fact be extremely sophisticated. Consider the extreme case where we have only one example of each class. For instance, we'd like to guess where the border between two countries is, but all we know is their capitals' locations. Most learners would be stumped, but nearest-neighbor happily guesses that the border is a straight line lying halfway between the two cities:

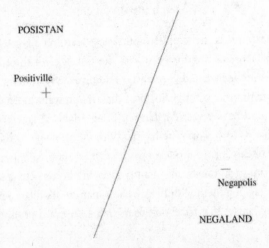

POSISTAN

Positiville
+

Negapolis

NEGALAND

The points on the line are at the same distance from the two capitals; points to the left of the line are closer to Positiville, so nearest-neighbor assumes they're part of Posistan and vice versa. Of course, it would be a lucky day if that was the exact border, but as an approximation it's probably a lot better than nothing. It's when we know a lot of towns on both sides of the border, though, that things get really interesting:

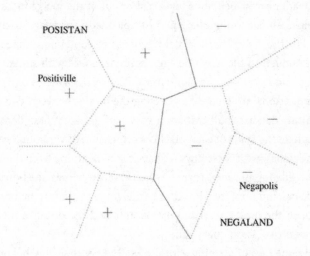

Nearest-neighbor is able to implicitly form a very intricate border, even though all it's doing is remembering where the towns are and assigning points to countries accordingly! We can think of the "metro area" of a town as all the points that are closer to it than to any other town; the boundaries between metro areas are shown as dashed lines in the diagram. Now Posistan is just the union of the metro areas of all its cities, as is Negaland. In contrast, a decision tree (for example) would only be able to form borders running alternately north–south and east–west, probably a much worse approximation to the real border. Thus, even though decision tree learners are "eager," trying hard at learning time to figure out where the border lies, "lazy" nearest-neighbor actually wins out.

The reason lazy learning wins is that forming a global model, such as a decision tree, is much harder than just figuring out where specific query points lie, one at a time. Imagine trying to define what a face is with a decision tree. You could say it has two eyes, a nose, and a mouth, but what is an eye and how do you find it in an image? What if the person's eyes are closed? Reliably defining a face all the way down to individual pixels is extremely difficult, particularly given all the different expressions, poses, contexts, and lighting conditions a face could appear in. Instead, nearest-neighbor takes a shortcut: if the image in its database most similar to the one Jane just uploaded is of a face, then so is Jane's. For this to work, the database needs to contain an image that's similar enough to the new one—for example, a face with similar pose, lighting, and so on—so the bigger the database, the better. For a simple two-dimensional problem like guessing the border between two countries, a tiny database suffices. For a very hard problem like identifying faces, where the color of each pixel is a dimension of variation, we need a huge database. But these days we have them. Learning from them may be too costly for an eager learner, which explicitly draws the border between faces and nonfaces. For nearest-neighbor, however, the border is implicit in the locations of the data points and the distance measure, and the only cost is at query time.

The same idea of forming a local model rather than a global one applies beyond classification. Scientists routinely use linear regression to predict continuous variables, but most phenomena are not linear. Luckily, they're locally linear because smooth curves are locally well approximated by straight lines. So if instead of trying to fit a straight line to all the data, you just fit it to the points near the query point, you now have a very powerful nonlinear regression algorithm. Laziness pays. If Kennedy had needed a complete theory of international relations to decide what to do about the Soviet missiles in Cuba, he would have been in trouble. Instead, he saw an analogy between that crisis and the outbreak of World War I, and that analogy guided him to the right decisions.

Nearest-neighbor can save lives, as Steven Johnson recounted in *The Ghost Map*. In 1854, London was struck by a cholera outbreak, which killed as many as one in eight people in parts of the city. The then-prevailing theory that cholera was caused by "bad air" did nothing to prevent its spread. But John Snow, a physician who was skeptical of the theory, had a better idea. He marked on a map of London the locations of all the known cases of cholera and divided the map into the regions closest to each public water pump. Eureka: nearly all deaths were in the "metro area" of one particular pump, located on Broad Street in the Soho district. Inferring that the water in that well was contaminated, Snow convinced the locals to disable the pump, and the epidemic died out. This episode gave birth to the science of epidemiology, but it's also the first success of the nearest-neighbor algorithm—almost a century before its official invention.

With nearest-neighbor, each data point is its own little classifier, predicting the class for all the query examples it wins. Nearest-neighbor is like an army of ants, in which each soldier by itself does little, but together they can move mountains. If an ant's load is too heavy, it can share it with its neighbors. In the same spirit, in the k-nearest-neighbor algorithm, a test example is classified by finding its k nearest neighbors and letting them vote. If the nearest image to the new upload is a face but the next two nearest ones aren't, three-nearest-neighbor decides that the new upload is not a face after all. Nearest-neighbor is prone to overfitting: if we have the wrong class for a data point, it spreads to its entire metro area. K-nearest-neighbor is more robust because it only goes wrong if a majority of the k nearest neighbors is noisy. The price, of course, is that its vision is blurrier: fine details of the frontier get washed away by the voting. When k goes up, variance decreases, but bias increases.

Using the k nearest neighbors instead of one is not the end of the story. Intuitively, the examples closest to the test example should count for more. This leads us to the weighted k-nearest-neighbor algorithm. In 1994, a team of researchers from the University of Minnesota and MIT built a recommendation system based on what they called "a deceptively simple idea": people who agreed in the past are likely to agree again in the future. That notion led directly to the collaborative filtering systems that all self-respecting e-commerce sites have. Suppose that, like Netflix, you've gathered a database of movie ratings, with each user giving a rating of one to five stars to the movies he or she has seen. You want to decide whether your user Ken will like *Gravity*, so you find the users whose past ratings correlate most highly with his. If they all gave *Gravity* high ratings, then probably so will Ken, and you can recommend it to him. If they disagree on *Gravity*, however, you need a fallback point, which in this case is ranking users by how highly they correlate with Ken. So if Lee's correlation with Ken is higher than Meg's, his ratings should count for correspondingly more. Ken's predicted rating is then the weighted average of his neighbors', with each neighbor's weight being his coefficient of correlation with Ken.

There's an interesting twist, though. Suppose Lee and Ken have very similar tastes, but Lee is grumpier than Ken. Whenever Ken gives a movie five stars, Lee gives three; when Ken gives three, Lee gives one, and so on. We'd like to use Lee's ratings to predict Ken's, but if we just do it directly, we'll always be off by two stars. Instead, what we need to do is predict how much Ken's ratings will be above or below his average, based on how much Lee's are. And now, since Ken is always two stars above his average when Lee is two stars above his, and so on, our predictions will be spot on.

You don't need explicit ratings to do collaborative filtering, by the way. If Ken ordered a movie on Netflix, that means he expects to like it. So the "ratings" can just be ordered/not ordered, and two users are similar if they've ordered a lot of the same movies. Even just clicking on something implicitly shows interest in it. Nearest-neighbor works with all of the above. These days all kinds of algorithms are used to recommend items to users, but weighted *k*-nearest-neighbor was the first widely used one, and it's still hard to beat.

Recommender systems, as they're also called, are big business: a third of Amazon's business comes from its recommendations, as does three-quarters of Netflix's. It's a far cry from the early days of nearest-neighbor, when it was considered impractical because of its memory requirements. Back then, computer memories were made of small iron rings, one per bit, and storing even a few thousand examples was taxing. How times have changed. Nevertheless, it's not necessarily smart to remember all the examples you've seen and then have to search through them, particularly since most are probably irrelevant. If you look back at the map of Posistan and Negaland, you may notice that if Positiville disappeared, nothing would change. The metro areas of nearby cities would expand into the land formerly occupied by Positiville, but since they're all Posistan cities, the border with Negaland would stay the same. The only cities that really matter are the ones across the border from a city in the other country; all others we can omit. So a simple way to make nearest-neighbor more efficient is to delete all the examples that are correctly classified by their neighbors. This and other tricks enable nearest-neighbor methods to be used in some surprising areas, like controlling robot arms in real time. But needless to say, they're still not the first choice for things like high-frequency trading, where computers buy and sell stocks in fractions of a second. In a race between a neural network, which can be applied to an example with only a fixed number of additions, multiplications, and sigmoids and an algorithm that needs to search a large database for the example's nearest neighbors, the neural network is sure to win.

Another reason researchers were initially skeptical of nearest-neighbor was that it wasn't clear if it could learn the true borders between concepts. But in 1967 Tom Cover and Peter Hart proved that, given enough data, nearest-neighbor is at worst only twice as error-prone as the best imaginable classifier. If, say, at least 1 percent of test examples will inevitably be misclassified because of noise in the data, then nearest-neighbor is guaranteed to get at most 2 percent wrong. This was a momentous revelation. Up until then, all known classifiers assumed that the frontier had a very specific form, typically a straight line. This was a double-edged sword: on the one hand, it made proofs of correctness possible, as in the case of the perceptron, but it also meant that the classifier was strictly limited in what it could learn. Nearest-neighbor was the first algorithm in history that could take advantage of unlimited amounts of data to learn arbitrarily complex concepts. No human being could hope to trace the frontiers it forms in hyperspace from millions of examples, but because of Cover and Hart's proof, we know that they're probably not far off the mark. According to Ray Kurzweil, the Singularity begins when we can no longer understand what computers do. By that standard, it's not entirely fanciful to say that it's already under way—it began all the way back in 1951, when Fix and Hodges invented nearest-neighbor, the little algorithm that could.

The curse of dimensionality

There's a serpent in this Eden, of course. It's called the curse of dimensionality, and while it affects all learners to a greater or lesser degree, it's particularly bad for nearest-neighbor. In low dimensions (like two or three), nearest-neighbor usually works quite well. But as the number of dimensions goes up, things fall apart pretty quickly. It's not uncommon today to have thousands or even millions of attributes to learn from. For an e-commerce site trying to learn your preferences, every click you make is an attribute. So is every word on a web page, and every pixel on an image. But even with just tens or hundreds of attributes, chances are nearest-neighbor is already in trouble. The first problem is that most attributes are irrelevant: you may know a million factoids about Ken, but chances are only a few of them have anything to say about (for example) his risk of getting lung cancer. And while knowing whether he smokes is crucial for making that particular prediction, it's probably not much help in deciding whether he'll enjoy seeing *Gravity*. Symbolist methods, for one, are fairly good at disposing of irrelevant attributes. If an attribute has no information about the class, it's just never included in the decision tree or rule set. But nearest-neighbor is hopelessly confused by irrelevant attributes because they all contribute to the similarity between examples. With enough irrelevant attributes, accidental similarity in the irrelevant dimensions swamps out meaningful similarity in the important ones, and nearest-neighbor becomes no better than random guessing.

A bigger problem is that, surprisingly, having more attributes can be harmful even when they're all relevant. You'd think that more information is always better—isn't that the motto of our age? But as the number of dimensions goes up, the number of training examples you need to locate the concept's frontiers goes up exponentially. With twenty Boolean attributes, there are roughly a million different possible examples. With twenty-one, there are two million, and a corresponding number of ways the frontier could wind between them. Every extra attribute makes the learning problem twice as hard, and that's just with Boolean attributes. If the attribute is highly informative, the benefit of adding it may exceed the cost. But if you have only weakly informative attributes, like the words in an e-mail or the pixels in an image, you're probably in trouble, even though collectively they may have enough information to predict what you want.

It gets even worse. Nearest-neighbor is based on finding similar objects, and in high dimensions, the notion of similarity itself breaks down. Hyperspace is like the Twilight Zone. The intuitions we have from living in three dimensions no longer apply, and weird and weirder things start to happen. Consider an orange: a tasty ball of pulp surrounded by a thin shell of skin. Let's say 90 percent of the radius of an orange is occupied by pulp, and the remaining 10 percent by skin. That means 73 percent of the volume of the orange is pulp (0.9^3). Now consider a hyperorange: still with 90 percent of the radius occupied by pulp, but in a hundred dimensions, say. The pulp has shrunk to only about three thousandths of a percent of the hyperorange's volume (0.9^{100}). The hyperorange is all skin, and you'll never be done peeling it!

Another disturbing example is what happens with our good old friend, the normal distribution, aka a bell curve. What a normal distribution says is that data is essentially located at a point (the mean of the distribution), but with some fuzz around it (given by the standard deviation). Right? Not in hyperspace. With a high-dimensional normal distribution, you're more likely to get a sample far from the mean than close to it. A bell curve in hyperspace looks more like a doughnut than a bell. And when nearest-neighbor walks into this topsy-turvy world, it gets hopelessly confused. All examples look equally alike, and at the same time they're too far from each other to make useful predictions. If you sprinkle examples uniformly at random inside a high-dimensional hypercube, most are closer to a face of the cube than to their nearest neighbor. In medieval maps, uncharted areas were marked with dragons, sea serpents, and other fantastical creatures, or just with the phrase *here be dragons*. In hyperspace, the dragons are everywhere, including at your front door. Try to walk to your next-door neighbor's house, and you'll never get there; you'll be forever lost in strange lands, wondering where all the familiar things went.

Decision trees are not immune to the curse of dimensionality either. Let's say the concept you're trying to learn is a sphere: points inside it are positive, and points outside it are negative. A decision tree can approximate a sphere by the smallest cube it fits inside. Not perfect, but not too bad either: only the corners of the cube get misclassified. But in high dimensions, almost the entire volume of the hypercube lies outside the hypersphere. For every example you correctly classify as positive, you incorrectly classify many negative ones as positive, causing your accuracy to plummet.

In fact, no learner is immune to the curse of dimensionality. It's the second worst problem in machine learning, after overfitting. The term *curse of dimensionality* was coined by Richard Bellman, a control theorist, in the fifties. He observed that control algorithms that worked fine in three dimensions became hopelessly inefficient in higher-dimensional spaces, such as when you want to control every joint in a robot arm or every knob in a chemical plant. But in machine learning the problem is more than just computational cost—it's that learning itself becomes harder and harder as the dimensionality goes up.

All is not lost, however. The first thing we can do is get rid of the irrelevant dimensions. Decision trees do this automatically by computing the information gain of each attribute and using only the most informative ones. For nearest-neighbor, we can accomplish something similar by first discarding all attributes whose information gain is below some threshold and then measuring similarity only in the reduced space. This is quick and good enough for some applications, but unfortunately it precludes learning many concepts, like exclusive-OR: if an attribute only says something about the class when combined with others, but not on its own, it will be discarded. A more expensive but smarter option is to "wrap" the attribute selection around the learner itself, with a hill-climbing search that keeps deleting attributes as long as that doesn't hurt nearest-neighbor's accuracy on held-out data. Newton did a lot of attribute selection when he decided that all that matters for predicting an object's trajectory is its mass—not its color, smell, age, or myriad other properties. In fact, the most important thing about an equation is all the quantities that don't appear in it: once we know what the essentials are, figuring out how they depend on each other is often the easier part.

To handle weakly relevant attributes, one option is to learn attribute weights. Instead of letting the similarity along all dimensions count equally, we "shrink" the less-relevant ones. Suppose the training examples are points in a room, and the height dimension is not that important for our purposes. Discarding it would project all examples onto the floor. Downweighting it is more like giving the room a lower ceiling. The height of a point still counts when computing its distance to other points, but less than its horizontal position. And like many other things in machine learning, we can learn attribute weights by gradient descent.

It may happen that the room has a high ceiling, but the data points are all near the floor, like a thin layer of dust settling on the carpet. In that case, we're in luck: the problem looks three dimensional, but in effect it's closer to two dimensional. We don't have to shrink height because nature has already shrunk it for us. This "blessing of nonuniformity," whereby data is not spread uniformly in (hyper) space, is often what saves the day. The examples may have a thousand attributes, but in reality they all "live" in a much lower-dimensional space. That's why nearest-neighbor can be good for handwritten digit recognition, for example: each pixel is a dimension, so there are many, but only a tiny fraction of all possible images are digits, and they all live together in a cozy little corner of hyperspace. The shape of the lower-dimensional space the data lives in may be quite capricious, however. For example, if a room has furniture in it, the dust doesn't just settle on the floor;

it settles on the tabletops, chair seats, bed covers, and whatnot. If we can figure out the approximate shape of the blanket of dust covering the room, then all we need is each point's coordinates on it. As we'll see in the next chapter, there's a whole subfield of machine learning dedicated to, so to speak, discovering blanket shapes by groping around in the darkness of hyperspace.

Snakes on a plane

Up until the mid-1990s, the most widely used analogical learner was nearest-neigbhor, but it was overshadowed by its more glamorous cousins from the other tribes. But then a new similarity-based algo-rithm burst onto the scene, sweeping all before it. In fact, you could say it was another "peace dividend" from the end of the Cold War. Support vector machines, or SVMs for short, were the brainchild of Vladimir Vapnik, a Soviet frequentist. Vapnik spent most of his career at the In-stitute of Control Sciences in Moscow, but in 1990, as the Soviet Union unraveled, he emigrated to the United States, where he joined the leg-endary Bell Labs. While in Russia, Vapnik had been mostly content to do theoretical, pencil-and-paper work, but the atmosphere at Bell Labs was different. Researchers were looking for practical results, and Vapnik finally decided to turn his ideas into an algorithm. Within a few years, he and his colleagues at Bell Labs had developed SVMs, and before long they were everywhere, setting new accuracy records left and right.

Superficially, an SVM looks a lot like weighted k-nearest-neighbor: the frontier between the positive and negative classes is defined by a set of examples and their weights, together with a similarity measure. A test example belongs to the positive class if, on average, it looks more like the positive examples than the negative ones. The average is weighted, and the SVM remembers only the key examples required to pin down the fron-tier. If you look back at the Posistan/Negaland example, once we throw away all the towns that aren't on the border, all that's left is this map:

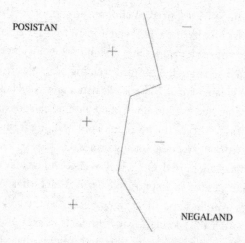

These examples are called support vectors because they're the vectors that "hold up" the frontier: remove one, and a section of the frontier slides to a different place. You may also notice that the frontier is a jagged line, with sudden corners that depend on the exact location of the examples. Real concepts tend to have smoother borders, which means nearest-neighbor's approximation is probably not ideal. But with SVMs, we can learn smooth frontiers, more like this:

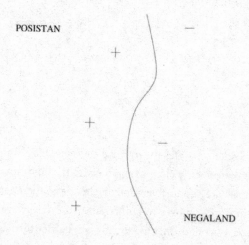

To learn an SVM, we need to choose the support vectors and their weights. The similarity measure, which in SVM-land is called the kernel, is usually chosen a priori. One of Vapnik's key insights was that not all borders that separate the positive training examples from the negative ones are created equal. Suppose Posistan and Negaland are at war, and they're separated by a no-man's-land with minefields on either side. Your mission is to survey the no-man's-land, walking from one end of it to the other without stepping on any mines. Luckily, you have a map of where the mines are buried. Obviously, you don't just take any old path: you give the mines the widest possible berth. That's what SVMs do, with the examples as mines and the learned border as the chosen path. The closest the border ever comes to an example is its margin of safety, and the SVM chooses the support vectors and weights that yield the maximum possible margin. For example, the solid straight-line border in this figure is better than the dotted one:

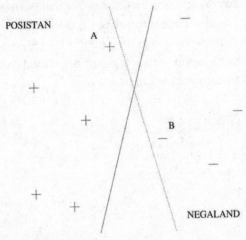

The dotted border separates the positive and negative examples just fine, but it comes dangerously close to stepping on the landmines at A and B. These examples are support vectors: delete one of them, and the maximum-margin border moves to a different place. In general, the border can be curved, of course, making the margin harder to visualize, but we can think of the border as a snake slithering down the no-man's-land,

and the margin is how fat the snake can be. If a very fat snake can slither all the way down without blowing itself to smithereens, then the SVM can separate the positive and negative examples very well, and Vapnik showed that in this case we can be confident that the SVM didn't overfit. Intuitively, compared to a thin snake, there are fewer ways a fat snake can slither down while avoiding the landmines; and likewise, compared to a low-margin SVM, a high-margin one has fewer chances of overfitting by drawing an overly intricate border.

The second part of the story is how the SVM finds the fattest snake that fits between the positive and negative landmines. At first sight, it might seem like learning a weight for each training example by gradient descent would do the trick. All we have to do is find the weights that maximize the margin, and any examples that end up with zero weight can be discarded. Unfortunately, this would just make the weights grow without limit, because mathematically, the larger the weights, the larger the margin. If you're one foot from a landmine and you double the size of everything including yourself, you are now two feet from the landmine, but that doesn't make you any less likely to step on it. Instead, we have to maximize the margin under the constraint that the weights can only increase up to some fixed value. Or, equivalently, we can minimize the weights under the constraint that all examples have a given margin, which could be one—the precise value is arbitrary. This is what SVMs usually do.

Constrained optimization is the problem of maximizing or minimizing a function subject to constraints. The universe maximizes entropy subject to keeping energy constant. Problems of this type are widespread in business and technology. For example, we may want to maximize the number of widgets a factory produces, subject to the number of machine tools available, the widgets' specs, and so on. With SVMs, constrained optimization became crucial for machine learning as well. Unconstrained optimization is getting to the top of the mountain, and that's what gradient descent (or, in this case, ascent) does. Constrained optimization is going as high as you can while staying on the road. If the road goes up to the very top, the constrained and unconstrained problems have the same solution. More often, though, the road zigzags up the mountain and then back down without ever reaching the top. You know you've reached the highest point on the road when you can't go any higher without driving off the road; in other words, when the path to the top is at right angles to the road. If the road and the path to the top form an oblique angle, you can always get higher by driving farther along the road, even if that doesn't get you higher as quickly as aiming straight for the top of the mountain. So the way to solve a constrained optimization problem is to follow not the gradient but the part of it that's parallel to the constraint surface—in this case the road—and stop when that part is zero.

In general, we have to deal with many constraints at once (one per example, in the case of SVMs). Suppose you wanted to get as close as possible to the North Pole but couldn't leave your room. Each of the room's four walls is a constraint, and the solution is to follow the compass until you bump into the corner where the northeast and northwest walls meet. We say that these two walls are the active constraints because they're what prevents you from reaching the optimum, namely the North Pole. If your room has a wall facing exactly north, that's the sole active constraint, and the solution is a point in the middle of it. And if you're Santa and your room is already over the North Pole, all constraints are inactive, and you can just sit there pondering the optimal toy distribution problem instead. (Traveling salesmen have it easy compared to Santa.) In an SVM, the active constraints are the support vectors since their margin is already the smallest it's allowed to be; moving the frontier would violate one or more constraints. All other examples are irrelevant, and their weight is zero.

In reality, we usually let SVMs violate some constraints, meaning classify some examples incorrectly or by less than the margin, because otherwise they would overfit. If there's a noisy negative example somewhere in the middle of the positive region, we don't want the frontier to wind around inside the positive region just to get that example right. But the SVM pays a penalty for each example it gets wrong, which encourages it to keep those to a minimum. SVMs are like the sandworms in *Dune*: big, tough, and able to survive a few explosions from slithering over landmines but not too many.

Looking around for applications, Vapnik and his coworkers soon alighted on handwritten digit recognition, which their connectionist colleagues at Bell Labs were the world experts on. To everyone's surprise, SVMs did as well out of the box as multilayer perceptrons that had been carefully crafted for digit recognition over the years. This set the stage for a long-running, wide-ranging competition between the two. SVMs can be seen as a generalization of the perceptron, because a hyperplane boundary between classes is what you get when you use a particular similarity measure (the dot product between vectors). But SVMs have a major advantage compared to multilayer perceptrons: the weights have a single optimum instead of many local ones and so learning them reliably is much easier. Despite this, SVMs are no less expressive than multilayer perceptrons; the support vectors effectively act as a hidden layer and their weighted average as the output layer. For example, an SVM can easily represent the exclusive-OR function by having one support vector for each of the four possible configurations. But the connectionists didn't give up without a fight. In 1995, Larry Jackel, the head of Vapnik's department at Bell Labs, bet him a fancy dinner that by 2000 neural networks would be as well understood as SVMs. He lost. But in return, Vapnik bet that by 2005 no one would use neural networks any more, and he also lost. (The only one to get a free dinner was Yann LeCun, their witness.) Moreover, with the advent of deep learning, connectionists have regained the upper hand. Provided you can learn them, networks with many layers can express many functions more compactly than SVMs, which always have just one layer, and this can make all the difference.

Another notable early success of SVMs was in text classification, which proved a major boon because the web was then just taking off. At the time, Naïve Bayes was the state-of-the-art text classifier, but when every word in the language is a dimension, even it can start to overfit. All it takes is a word that, by chance, occurs in, say, all sports pages in the training data and no others, and Naïve Bayes starts to hallucinate that every page containing that word is a sports page. But, thanks to margin maximization, SVMs can resist overfitting even in very high dimensions.

Generally, the fewer support vectors an SVM selects, the better it generalizes. Any training example that is not a support vector would be correctly classified if it showed up as a test example instead because the frontier between positive and negative examples would still be in the same place. So the expected error rate of an SVM is at most the fraction of examples that are support vectors. As the number of dimensions goes up, this fraction tends to go up as well, so SVMs are not immune to the curse of dimensionality. But they're more resistant to it than most.

Practical successes aside, SVMs also turned a lot of machine-learning conventional wisdom on its head. For example, they gave the lie to the notion, sometimes misidentified with Occam's razor, that simpler models are more accurate. On the contrary, an SVM can have an infinite number of parameters and still not overfit, provided it has a large enough margin.

The single most surprising property of SVMs, however, is that no matter how curvy the frontiers they form, those frontiers are always just straight lines (or hyperplanes, in general). The reason that's not a contradiction is that the straight lines are in a different space. Suppose the examples live on the (x,y) plane, and the boundary between the positive and negative regions is the parabola $y = x^2$. There's no way to represent it with a straight line, but if we add a third coordinate z, meaning the data now lives in (x,y,z) space, and we set each example's z coordinate to the square of its x coordinate, the frontier is now just the diagonal plane defined by $y = z$. In effect, the data points rise up into the third dimension, some rise more than others by just the right amount, and presto—in this new dimension the positive and negative examples can be separated by a plane. It turns out that we can view what SVMs do with kernels, support vectors, and weights as mapping the data to a higher-dimensional space and finding a maximum-margin hyperplane in that space. For some kernels, the derived space has infinite dimensions, but SVMs are completely unfazed by that. Hyperspace may be the Twilight Zone, but SVMs have figured out how to navigate it.

Climbing the ladder

Two things are similar if they agree with one another in some respects. If they agree in some respects, they will probably also agree in others. This is the essence of analogy. It also points to the two main subproblems in analogical reasoning: figuring out how similar two things are and deciding what else to infer from their similarities. So far we've explored the "low power" end of analogy, with algorithms like nearest-neighbor and SVMs, where the answers to both these questions are very simple. They're the most widely used, but a chapter on analogical learning would not be complete without at least a whirlwind tour of the more powerful parts of the spectrum.

The most important question in any analogical learner is how to measure similarity. It could be as simple as Euclidean distance between data points, or as complex as a whole program with multiple levels of subroutines whose final output is a similarity value. Either way, the similarity function controls how the learner generalizes from known examples to new ones. It's where we insert our knowledge of the problem domain into the learner, making it the analogizers' answer to Hume's question. We can apply analogical learning to all kinds of objects, not just vectors of attributes, provided we have a way of measuring the similarity between them. For example, we can measure the similarity between two molecules by the number of identical substructures they contain. Methane and methanol are similar because they have three carbon-hydrogen bonds in common and differ only in the replacement of a hydrogen atom by a hydroxyl group:

Methane   Methanol

However, that doesn't mean their chemical behavior is similar. Methane is a gas, while methanol is an alcohol. The second part of analogical reasoning is figuring out what we can infer about the new object based on similar ones we've found. This can be very simple or very complex. In nearest-neighbor or SVMs, it just consists of predicting the new object's class based on the classes of the nearest neighbors or support vectors. But in case-based reasoning, another type of analogical learning, the output can be a complex structure formed by composing parts of the retrieved objects. Suppose your HP printer is spewing out gibberish, and you call up their help desk. Chances are they've seen your problem many times before, so a good strategy is to find those records and piece together a potential solution for your problem from them. This is not just a matter of finding complaints with many similar attributes to yours: for example, whether you're using your printer with Windows or Mac OS X may cause very different settings of the system and the printer to become relevant. And once you've found the most relevant cases, the sequence of steps needed to solve your problem may be a combination of steps from different cases, with some further tweaks specific to yours.

Help desks are currently the most popular application of case-based reasoning. Most still employ a human intermediary, but IPsoft's Eliza talks directly to the customer. Eliza, who comes complete with a 3-D interactive video persona, has solved over twenty million customer problems to date, mostly for blue-chip US companies. "Greetings from Robotistan, outsourcing's cheapest new destination," is how an outsourcing blog recently put it. And, just as outsourcing keeps climbing the skills ladder, so does analogical learning. The first robo-lawyers that argue for a particular verdict based on precedents have already been built. One such system correctly predicted the outcomes of over 90 percent of the trade secret cases it examined. Perhaps in a future cybercourt, in session somewhere on Amazon's cloud, a robo-lawyer will beat the speeding ticket that RoboCop issued to your driverless car, all while you go to the beach, and Leibniz's dream of reducing all argument to calculation will finally have come true.

Arguably even higher up in the skills ladder is music composition. David Cope, an emeritus professor of music at the University of California, Santa Cruz, designed an algorithm that creates new music in the style of famous composers by selecting and recombining short passages from their work. At a conference I attended some years ago, he played three "Mozart" pieces: one by the real Mozart, one by a human composer imitating Mozart, and one by his system. He then asked the audience to vote for the authentic Amadeus. Wolfgang won, but the computer beat the human imitator. This being an AI conference, the audience was delighted. Audiences at other events were less happy. One listener angrily accused Cope of ruining music for him. If Cope is right, creativity—the ultimate unfathomable—boils down to analogy and recombination. Judge for yourself by googling "david cope mp3."

Analogizers' neatest trick, however, is learning across problem domains. Humans do it all the time: an executive can move from, say, a media company to a consumer-products one without starting from scratch because many of the same management skills still apply. Wall Street hires lots of physicists because physical and financial problems, although superficially very different, often have a similar mathematical structure. Yet all the learners we've seen so far would fall flat if we, say, trained them to predict Brownian motion and then asked them to predict the stock market. Stock prices and the velocities of particles suspended in a fluid are just different variables, so the learner wouldn't even know where to start. But analogizers can do this using structure mapping, an algorithm invented by Dedre Gentner, a psychologist at Northwestern University. Structure mapping takes two descriptions, finds a coherent correspondence between some of their parts and relations, and then, based on that correspondence, transfers further properties from one structure to the other. For example, if the structures are the solar system and the atom, we can map planets to electrons and the sun to the nucleus and conclude, as Bohr did, that electrons revolve around the nucleus. The truth is more subtle, of course, and we often need to refine analogies after we make them. But being able to learn from a single example like this is surely a key attribute of a universal learner. When we're confronted with a new type of cancer—and that happens all the time because cancers keep mutating—the models we've learned for previous ones don't apply. Neither do we have time to gather data on the new cancer from a lot of patients; there may be only one, and she urgently needs a cure. Our best hope is then to compare the new cancer with known ones and try to find one whose behavior is similar enough that some of the same lines of attack will work.

Is there anything analogy can't do? Not according to Douglas Hofstadter, cognitive scientist and author of *Gödel, Escher, Bach: An Eternal Golden Braid*. Hofstadter, who looks a bit like the Grinch's good twin, is probably the world's best-known analogizer. In their book *Surfaces and Essences: Analogy as the Fuel and Fire of Thinking*, Hofstadter and his collaborator Emmanuel Sander argue passionately that all intelligent behavior reduces to analogy. Everything we learn or discover, from the meaning of everyday words like *mother* and *play* to the brilliant insights of geniuses like Albert Einstein and Évariste Galois, is the result of analogy in action. When little Tim sees women looking after other children like his mother looks after him, he generalizes the concept "mommy" to mean anyone's mommy, not just his. That in turn is a springboard for understanding things like "mother ship" and "Mother Nature." Einstein's "happiest thought," out of which grew the general theory of relativity, was an analogy between gravity and acceleration: if you're in an elevator, you can't tell whether your weight is due to one or the other because their effects are the same. We swim in a vast ocean of analogies, which we both manipulate for our ends and are unwittingly manipulated by. Books have analogies on every page (like the title of this section, or the previous one's). *Gödel, Escher, Bach* is an extended analogy between Gödel's theorem, Escher's art, and Bach's music. If the Master Algorithm is not analogy, it must surely be something like it.

Rise and shine

Cognitive science has seen a long-running debate between symbolists and analogizers. Symbolists point to something they can model that analogizers can't; then analogizers figure out how to do it, come up with something they can model that symbolists can't, and the cycle repeats. Instance-based learning, as it's sometimes called, is supposedly better for modeling how we remember specific episodes in our lives; rules are the putative choice for reasoning with abstract concepts like "work" and "love." But when I was a graduate student, it struck me that these two are really just points on a continuum, and we should be able to learn across all of it. Rules are in effect generalized instances where we've "forgotten" some attributes because they didn't matter. Conversely, instances are very specific rules, with a condition on every attribute. As we go through life, similar episodes gradually become abstracted into rule-based structures, like "eating at a restaurant." You know that going to a restaurant involves ordering from a menu and leaving a tip, and you follow those "rules of conduct" every time you eat out, but you probably don't remember the specific restaurants where you first became aware of them.

In my PhD thesis, I designed an algorithm that unifies instance-based and rule-based learning in this way. A rule doesn't just match entities that satisfy all its preconditions; it matches any entity that's more similar to it than to any other rule, in the sense that it comes closer to satisfying its conditions. For instance, someone with a cholesterol level of 220 mg/ dL comes closer than someone with 200 mg/dL to matching the rule *If your cholesterol is above 240 mg/dL, you're at risk of a heart attack.* RISE, as I called the algorithm, learns by starting with each training example as a rule and then gradually generalizing each rule to absorb the nearest examples. The end result is usually a combination of very general rules, which between them match most examples, with more specific rules that match exceptions to those, and so on all the way to a "long tail" of specific memories. RISE made better predictions than the best rule-based and instance-based learners of the time, and my experiments showed that this was precisely because it combined the best features of both. Rules can be matched analogically, and so they're no longer brittle. Instances can select different features in different regions of space and so combat the curse of dimensionality much better than nearest-neighbor, which can only select the same features everywhere.

RISE was a step toward the Master Algorithm because it combined symbolic and analogical learning. It was only a small step, however, because it doesn't have the full power of either of those paradigms, and it's still missing the other three. RISE's rules can't be chained together in different ways; each rule just predicts the class of an example directly from its attributes. Also, the rules can't talk about more than one entity at a time; for example, RISE can't express a rule like *If A has the flu and B was in contact with A, B may have the flu as well.* On the analogical side, RISE just generalizes the simple nearest-neighbor algorithm; it can't learn across domains using structure mapping or some such strategy. At the time I finished my PhD, I didn't see a way to bring together in one algorithm the full power of all the five paradigms, and I set the problem aside for a while. But as I applied machine learning to problems like word-of-mouth marketing, data integration, programming by example, and website personalization, I kept seeing how each of the paradigms provided only part of the solution. There had to be a better way.

And so we have traveled through the territories of the five tribes, gathering their insights, negotiating the border crossings, wondering how the pieces might fit together. We know immensely more now than when we started out. But something is still missing. There's a gaping hole in the center of the puzzle, making it hard to see the pattern. The problem is that all the learners we've seen so far need a teacher to tell them the right answer. They can't learn to distinguish tumor cells from healthy ones unless someone labels them "tumor" or "healthy." But humans can learn without a teacher; they do it from the day they're born. Like Frodo at the gates of Mordor, our long journey will have been in vain if we don't find a way around this barrier. But there is a path past the ramparts and the guards, and the prize is near. Follow me . . .

CHAPTER EIGHT

Learning Without a Teacher

If you're a parent, the entire mystery of learning unfolds before your eyes in the first three years of your child's life. A newborn baby can't talk, walk, recognize objects, or even understand that an object continues to exist when the baby isn't looking at it. But month after month, in steps large and small, by trial and error and great conceptual leaps, the child figures out how the world works, how people behave, and how to communicate. By a child's third birthday, all this learning has coalesced into a stable self, a stream of consciousness that will continue throughout life. Older children and adults can time-travel, aka remember things past, but only so far back. If we could revisit ourselves as infants and toddlers and see the world again through those newborn eyes, much of what puzzles us about learning—even about existence itself—would suddenly seem obvious. But as it is, the greatest mystery in the universe is not how it begins or ends, or what infinitesimal threads it's woven from, it's what goes on in a small child's mind: how a pound of gray jelly can grow into the seat of consciousness.

The scientific study of children's learning is still young, having begun in earnest only a few decades ago, but it has already come remarkably far. Infants can't answer questionnaires or follow experimental protocols, but we can infer a surprising amount about what goes on in their minds by videotaping and studying their reactions during experiments. A coherent picture emerges: an infant's mind isn't just the unfolding of a predefined genetic program or a biological device for recording correlations in sense data; rather, the infant's mind actively synthesizes his or her reality, and this reality changes quite radically over time.

Increasingly, and most relevant to us, cognitive scientists express their theories of children's learning in the form of algorithms. Many machine-learning researchers take inspiration from this. Everything we need is right there in a child's mind, if only we can somehow capture its essence in computer code. Some researchers even argue that the way to create intelligent machines is to build a robot baby and let him experience the world as a human baby does. We, the researchers, would be his parents (perhaps even with an assist from crowdsourcing, giving a whole new meaning to the term *global village*). Little Robby— let's call him that, in honor of the chubby but much taller robot in *Forbidden Planet*—is the only robot baby we'll ever have to build. Once he has learned everything a three-year-old knows, the AI problem is solved. We can copy the contents of his mind into as many other robots as we like, and they'll take it from there, the hardest part already accomplished.

The question, of course, is what algorithm should be running in Robby's brain at birth. Researchers influenced by child psychology look askance at neural networks because the microscopic workings of a neuron seem a million miles from the sophistication of even a child's most basic behaviors, like reaching for an object, grasping it, and inspecting it with wide, curious eyes. We need to model the child's learning at a higher level of abstraction, lest we miss the planet for the trees. Above all, even though children certainly get plenty of help from their parents, they learn mostly on their own, without supervision, and that's what seems most miraculous. None of the algorithms we've seen so far can do it, but we're about to see several that can—bringing us one step closer to the Master Algorithm.

Putting together birds of a feather

We flip the "on" switch, and Robby's video eyes open for the very first time. At once he's flooded with what William James memorably called the "blooming, buzzing confusion" of the world. With new images streaming in at a rate of dozens per second, one of the first things he must do is learn to organize them into larger chunks. The real world is made up of objects that persist over time, not random pixels changing arbitrarily from one moment to the next. Mommy isn't replaced by a smaller Mommy when she walks away. Putting a dish on the table doesn't make a white hole in it. A young baby is not surprised if a teddy bear passes behind a screen and reemerges as an airplane, but a one-year-old is. Somehow, he's figured out that teddy bears are different from airplanes and don't spontaneously transmute. Soon afterward, he'll figure out that some objects are more alike than others and start forming categories. Given a pile of toy horses and pencils to play with, a nine-month-old doesn't think to sort them into separate piles of horses and pencils, but an eighteen-month-old does.

Organizing the world into objects and categories is second nature to an adult but not to an infant, and even less to Robby the robot. We could endow him with a visual cortex in the form of a multilayer perceptron and show him labeled examples of all the objects and categories in the world—here's Mommy close up, here's Mommy far away—but we'd never be done. What we need is an algorithm that will spontaneously group together similar objects, or different images of the same object. This is the problem of clustering, and it's one of the most intensively studied in machine learning.

A cluster is a set of similar entities, or at a minimum, a set of entities that are more similar to each other than to members of other clusters. It's human nature to cluster things, and it's often the first step on the road to knowledge. When we look up at the night sky, we can't help seeing clusters of stars, and then we fancifully name them after shapes they resemble. Noticing that certain sets of elements had very similar chemical properties was the first step in discovering the periodic table. Each of those sets is now a column in it. Everything we perceive is a cluster, from friends' faces to speech sounds. Without them, we'd be lost: children can't learn a language before they learn to identify the characteristic sounds it's made of, which they do in their first year of life, and all the words they then learn mean nothing without the clusters of real things they refer to. Confronted with big data—a very large number of objects—our first recourse is to group them into a more manageable number of clusters. A whole market is too coarse, and individual customers are too fine, so marketers divide markets into segments, which is their word for clusters. Even objects themselves are at bottom clusters of their observations, from all the different angles light falls on Mommy's face to all the different sound waves baby hears as the word *mommy*. And we can't think without objects, which is perhaps why quantum mechanics is so unintuitive: we want to visualize the subatomic world as particles colliding, or waves interfering, but it's not really either.

We can represent a cluster by its prototypical element: the image of your mother that you see with your mind's eye or the quintessential cat, sports car, country house, or tropical beach. Peoria, Illinois, is the average American town, according to marketing lore. Bob Burns, a fifty-three-year-old building maintenance supervisor in Windham, Connecticut, is America's most ordinary citizen—at least if you believe Kevin O'Keefe's book *The Average American*. Anything described by numeric attributes—say, people's heights, weights, girths, shoe sizes, hair lengths, and so on—makes it easy to compute the average member: his height is the average height of all the cluster members, his weight the average of all the weights, and so on. For categorical attributes, like gender, hair color, zip code, or favorite sport, the "average" is simply the most frequent value. The average member described by this set of attributes may or may not be a real person, but either way it's a useful reference to have: if you're brainstorming how to market a new product, picturing Peoria as the town where you're launching it or Bob Burns as your target customer beats thinking of abstract entities like "the market" or "the consumer."

As useful as such averages are, we can do even better; indeed the whole point of big data and machine learning is to avoid thinking at such a coarse level. Our clusters can be very specialized sets of people or even different aspects of the same person: Alice buying books for work, for leisure, or as Christmas presents; Alice in a good mood versus Alice with the blues. Amazon would like to distinguish the books Alice buys for herself from the ones she buys for her boyfriend, as this would allow it to make appropriate recommendations at appropriate times. Unfortunately, purchases don't come labeled with "self-gift" or "for Bob," and Amazon needs to figure out how to group them.

Suppose the entities in Robby's world fall into five clusters (people, furniture, toys, food, and animals), but we don't know which things belong to which clusters. This is the type of problem that Robby faces when we switch him on. One simple option for sorting entities into clusters is to pick five random objects as the cluster prototypes and then compare each entity with each prototype and assign it to the most similar prototype's cluster. (As in analogical learning, the choice of similarity measure is important. If the attributes are numeric, it can be as simple as Euclidean distance, but there are many other options.) We now need to update the prototypes. After all, a cluster's prototype is supposed to be the average of its members, and although that was necessarily the case when each cluster had only one member, it generally won't be after we have added a bunch of new members to each cluster. So for each cluster, we compute the average properties of its members and make that the new prototype. At this point, we need to update the cluster memberships again: since the prototypes have moved, the closest prototype to a given entity may also have changed. Let's imagine the prototype of one category was a teddy bear and the prototype of another was a banana. Perhaps on our first run we grouped an animal cracker with the bear, but on the second we grouped it with the banana. An animal cracker initially looked like a toy, but now it looks more like food. Once I reclassify animal crackers in the banana group, perhaps the prototypical item for that group also changes, from a banana to a cookie. This virtuous cycle, with entities assigned to better and better clusters, continues until the assignment of entities to clusters doesn't change (and therefore neither do the cluster prototypes).

This algorithm is called k-means, and its origins go back to the fifties. It's nice and simple and quite popular, but it has several shortcomings, some of which are easier to solve than others. For one, we need to fix the number of clusters in advance, but in the real world, Robby is always running into new kinds of objects. One option is to let an object start a new cluster if it's too different from the existing ones. Another is to allow clusters to split and merge as we go along. Either way, we probably want the algorithm to include a preference for fewer clusters, lest we wind up with each object as its own cluster (hard to beat if we want clusters to consist of similar objects, but clearly not the goal).

A bigger issue is that k-means only works if the clusters are easy to tell apart: each cluster is roughly a spherical blob in hyperspace, the blobs are far from each other, and they all have similar volumes and include a similar number of objects. If any of these fails, ugly things can happen: an elongated cluster is split into two different ones, a smaller cluster is absorbed into a larger one nearby, and so on. Luckily, there's a better option.

Suppose we decide that letting Robby roam around in the real world is too slow and cumbersome a way to learn. Instead, like a would-be pilot learning in a flight simulator, we'll have him look at computer-generated images. We know what clusters the images come from, but we're not telling Robby. Instead, we create each image by first choosing a cluster at random (toys, say) and then synthesizing an example of that cluster (small, fluffy, brown teddy bear with big black eyes, round ears, and a bow tie). We also choose the properties of the example at random: the size comes from a normal distribution with a mean of ten inches, the fur is brown with 80 percent probability and white otherwise, and so on. After Robby has seen lots of images generated in this way, he should have learned to cluster them into people, furniture, toys, and so on, because people are more like people than furniture and so on. But the interesting question is: If we look at it from Robby's point of view, what's the best algorithm to discover the clusters? The answer is surprising: Naïve Bayes, which we first met as an algorithm for supervised learning. The difference is that now Robby doesn't know the classes, so he'll have to guess them!

Clearly, if Robby did know them, it would be smooth sailing: as in Naïve Bayes, each cluster would be defined by its probability (17 percent of the objects generated were toys), and by the probability distribution of each attribute among the cluster's members (for example, 80 percent of the toys are brown). Robby could estimate these probabilities just by counting the number of toys in the data, the number of brown toys, and so on. But in order to do that, we would need to know which objects are toys. This seems like a tough nut to crack, but it turns out we already know how to do it as well. If Robby has a Naïve Bayes classifier and needs to figure out the class of a new object, all he needs to do is apply the classifier and compute the probability of each class given the object's attributes. (Small, fluffy, brown, bear-like, with big eyes, and a bow tie? Probably a toy but possibly an animal.)

So Robby is faced with a chicken-and-egg problem: if he knew the objects' classes, he could learn the classes' models by counting, and if he knew the models, he could infer the objects' classes. We seem to be stuck again, but far from it: just start by guessing a class for each object any way you want—even at random—and you're off to the races. From those classes and the data, you can learn the class models; based on these models you can reinfer the classes and so on. At first sight this looks like a crazy scheme: it may never finish, circling forever between inferring the classes from the models and the models from the classes, and even if it does finish, there's no reason to believe it will settle on meaningful clusters. But in 1977 a trio of Harvard statisticians (Arthur Dempster, Nan Laird, and Donald Rubin) showed that the crazy scheme actually works: every time we go around the loop, the cluster model gets better, and the loop ends when the model is a local maximum of the likelihood. They called this scheme the EM algorithm, where the E stands for expectation (inferring the expected probabilities) and the M for maximization (estimating the maximum-likelihood parameters). They also showed that many previous algorithms were special cases of EM. For example, to learn hidden Markov models, we alternate between inferring the hidden states and estimating the transition and observation probabilities based on them. Whenever we want to learn a statistical model but are missing some crucial information (e.g., the classes of the examples), we can use EM. This makes it one of the most popular algorithms in all of machine learning.

You might have noticed a certain resemblance between k-means and EM, in that they both alternate between assigning entities to clusters and updating the clusters' descriptions. This is not an accident: k-means itself is a special case of EM, which you get when all the attributes have "narrow" normal distributions, that is, normal distributions with very small variance. When clusters overlap a lot, an entity could belong to, say, cluster A with a probability of 0.7 and cluster B with a probability of 0.3, and we can't just decide that it belongs to cluster A without losing information. EM takes this into account by fractionally assigning the entity to the two clusters and updating their descriptions accordingly. If the distributions are very concentrated, however, the probability that an entity belongs to the nearest cluster is always approximately 1, and all we have to do is assign entities to clusters and average the entities in each cluster to obtain its mean, which is just the k-means algorithm.

So far we've only seen how to learn one level of clusters, but the world is, of course, much richer than that, with clusters within clusters all the way down to individual objects: living things cluster into plants and animals, animals into mammals, birds, fishes, and so on, all the way down to Fido the family dog. No problem: once we've learned one set of clusters, we can treat them as objects and cluster them in turn, and so on up to the cluster of all things. Alternatively, we can start with a coarse clustering and then further divide each cluster into subclusters: Robby's toys divide into stuffed animals, constructions toys, and so on; stuffed animals into teddy bears, plush kittens, and so on. Children seem to start out in the middle and then work their way up and down. For example, they learn *dog* before they learn *animal* or *beagle*. This might be a good strategy for Robby, as well.

Discovering the shape of the data

Whether it's data pouring into Robby's brain through his senses or the click streams of millions of Amazon customers, grouping a large number of entities into a smaller number of clusters is only half the battle. The other half is shortening the description of each entity. The very first picture of Mom that Robby sees comprises perhaps a million pixels, each with its own color, but you hardly need a million variables to describe a face. Likewise, each thing you click on at Amazon provides an atom of information about you, but what Amazon would really like to know is your likes and dislikes, not your clicks. The former, which are fairly stable, are somehow immanent in the latter, which grow without limit as you use the site. Little by little, all those clicks should add up to a picture of your taste, in the same way that all those pixels add up to a picture of your face. The question is how to do the adding.

A face has only about fifty muscles, so fifty numbers should suffice to describe all possible expressions, with plenty of room to spare. The shape of the eyes, nose, mouth, and so on—the features that let you tell one person from another—shouldn't take more than a few dozen numbers, either. After all, with only ten choices for each facial feature, a police artist can put together a sketch of a suspect that's good enough to recognize him. You can add a few more numbers to specify lighting and pose, but that's about it. So if you give me a hundred numbers or so, that should be enough to re-create a picture of a face. Conversely, Robby's brain should be able to take in a picture of a face and quickly reduce it to the hundred numbers that really matter.

Machine learners call this process dimensionality reduction because it reduces a large number of visible dimensions (the pixels) to a few implicit ones (expression, facial features). Dimensionality reduction is essential for coping with big data—like the data coming in through your senses every second. A picture may be worth a thousand words, but it's also a million times more costly to process and remember. Yet somehow your visual cortex does a pretty good job of whittling it down to a manageable amount of information, enough to navigate the world, recognize people and things, and remember what you saw. It's one of the great miracles of cognition and so natural you're not even conscious of doing it.

When you arrange books on a shelf so that books on similar topics are close to each other, you're doing a kind of dimensionality reduction, from the vast space of topics to the one-dimensional shelf. Unavoidably, some books that are closely related will wind up far apart on the shelf, but you can still order them in a way that minimizes such occurrences. That's what dimensionality reduction algorithms do.

Suppose I give you the GPS coordinates of all the shops in Palo Alto, California, and you plot a few of them on a piece of paper:

You can probably tell just by looking at this plot that the main street in Palo Alto runs southwest–northeast. You didn't draw a street, but you can intuit that it's there from the fact that all the points fall along a straight line (or close to it—they can be on different sides of the street). Indeed, the street is University Avenue, and if you want to shop or eat out in Palo Alto, that's the place to go. As a bonus, once you know that the shops are on University Avenue, you don't need two numbers to locate them, just one: the street number (or, if you wanted to be really precise, the distance from the shop to the Caltrain station, on the southwest corner, which is where University Avenue begins).

If you plot more shops, you'll probably notice that some are on cross streets, a little bit off University Avenue, and a few are elsewhere entirely:

Nevertheless, it's still the case that most shops are pretty close to University Avenue, and if you were allowed only one number to locate a shop, its distance from the Caltrain station along the avenue would be a pretty good choice: after walking that distance, looking around is probably enough to find the shop. So you've just reduced the dimensionality of "shop locations in Palo Alto" from two to one.

Robby doesn't have the benefit of your highly evolved visual system, though, so if you want him to go fetch your dry cleaning from Elite Cleaners and you only allow his map of Palo Alto to have one coordinate, he needs an algorithm to "discover" University Avenue from the GPS coordinates of the shops. The key to this is to notice that, if you put the origin of the x,y plane at the average of the shops' locations and slowly rotate the axes, the shops are closest to the x axis when you've turned it by about 60 degrees, that is, when it lines up with University Avenue:

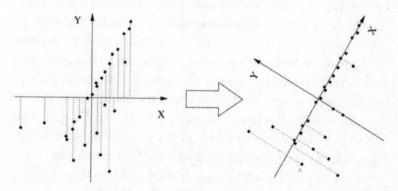

This direction—known as the first principal component of the data—is also the direction along which the spread of the data is greatest. (Notice how, if you project the shops onto the x axis, they're farther apart in the right figure than in the left one.) After you've found the first principal component, you can look for the second one, which in this case is the direction of greatest variation at right angles to University Avenue. On a map, there's only one possible direction left (the direction of the cross streets). But if Palo Alto was on a hillside, one or both of the two first principal components would be partly uphill, and the third and last one would be up into the air. We can apply the same idea to data in thousands or millions of dimensions, like face images, successively looking for the directions of greatest variation until the remaining variability is small, at which point we can stop. For example, after rotating the axes in the figure above, most shops have $y = 0$, so the average y is very small, and we don't lose too much information by ignoring the y coordinate altogether. And if we decide to keep y, surely z (up into the air) is insignificant. As it turns out, the whole process of finding the principal components can all be accomplished in one shot with a bit of linear algebra. Best of all, a few dimensions often account for the bulk of the variation in even very high-dimensional data. Even if that's not the case, eyeballing the data in the top two or three dimensions often yields a lot of insight because it takes advantage of your visual system's amazing powers of perception.

Principal-component analysis (PCA), as this process is known, is one of the key tools in the scientist's toolkit. You could say PCA is to unsupervised learning what linear regression is to the supervised variety. The famous hockey-stick curve of global warming, for example, is the result of finding the principal component of various temperature-related data series (tree rings, ice cores, etc.) and assuming it's the temperature. Biologists use PCA to summarize the expression levels of thousands of different genes into a few pathways. Psychologists have found that personality boils down to five dimensions—extroversion, agreeableness, conscientiousness, neuroticism, and openness to experience—which they can infer from your tweets and blog posts. (Chimps supposedly have one more dimension—reactivity—but Twitter data for them is not available.) Applying PCA to congressional votes and poll data shows that, contrary to popular belief, politics is not mainly about liberals versus conservatives. Rather, people differ along two main dimensions: one for economic issues and one for social ones. Collapsing these into a single axis mixes together populists and libertarians, who are polar opposites, and creates the illusion of lots of moderates in the middle. Trying to appeal to them is an unlikely winning strategy. On the other hand, if liberals and libertarians overcame their mutual aversion, they could ally themselves on social issues, where both favor individual freedom.

When he grows up, Robby can use a variant of PCA to solve the "cocktail party" problem, which is to pick out individual voices from the babble of the crowd. A related method can help him learn to read. If each word is a dimension, then a text is a point in the space of words, and the main directions of that space turn out to be elements of meaning. For example, *President Obama* and *the White House* are far apart in word space but close together in meaning space, because they tend to appear in similar contexts. Believe it or not, this type of analysis is all it takes for computers to grade SAT essays as well as humans do. Netflix uses a similar idea. Instead of just recommending movies that users with similar tastes liked, it first projects both users and movies into a lower-dimensional "taste space" and recommends a movie if it's close to you in this space. That way it can find movies for you that you never knew you'd love.

You'd probably be disappointed if you looked at the principal components of a face data set, though. They're not what you'd expect, such as facial expressions or features, but more like ghostly faces, blurred beyond recognition. This is because PCA is a linear algorithm, and so all that the principal components can be is weighted pixel-by-pixel averages of real faces. (Also known as eigenfaces because they're eigenvectors of the centered covariance matrix of the data—but I digress.) To really understand faces, and most shapes in the world, we need something else: nonlinear dimensionality reduction.

Suppose we zoom out from Palo Alto, and I give you the GPS coordinates of the main cities in the Bay Area:

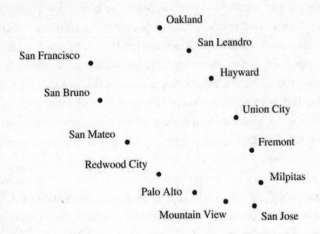

Again, you can probably surmise just by looking at this plot that the cities are on a bay, and if you draw a line running through them, you can locate each city using just one number: how far it is from San Francisco along that line. But PCA can't find this curve; instead, it draws a straight line running down the middle of the bay, where there are no cities at all. Far from elucidating the shape of the data, PCA obscures it.

Instead, imagine for a moment that we're going to develop the Bay Area from scratch. We've decided where each city will be located, and our budget allows us to build a single road connecting them. Naturally, we lay down a road that goes from San Francisco to San Bruno, from there to San Mateo, and so on all the way to Oakland. This road is a pretty good one-dimensional representation of the Bay Area and can be found by a simple algorithm: build a road between each pair of nearby cities. Of course, in general this will result in a network of roads, not a single road running by every city. But we can force the latter by building the single road that best approximates the network, in the sense that the distances between cities along this road are as close as possible to the distances along the network.

One of the most popular algorithms for nonlinear dimensionality reduction, called Isomap, does just this. It connects each data point in a high-dimensional space (a face, say) to all nearby points (very similar faces), computes the shortest distances between all pairs of points along the resulting network and finds the reduced coordinates that best approximate these distances. In contrast to PCA, faces' coordinates in this space are often quite meaningful: one may represent which direction the face is facing (left profile, three quarters, head on, etc.); another how the face looks (very sad, a little sad, neutral, happy, very happy, etc.); and so on. From understanding motion in video to detecting emotion in speech, Isomap has a surprising ability to zero in on the most important dimensions of complex data.

Here's an interesting experiment. Take the video stream from Robby's eyes, treat each frame as a point in the space of images, and reduce that set of images to a single dimension. What will you discover? Time. Like a librarian arranging books on a shelf, time places each image next to its most similar ones. Perhaps our perception of it is just a natural result of our brains' dimensionality reduction prowess. In the road network of memory, time is the main thoroughfare, and we soon find it. Time, in other words, is the principal component of memory.

The hedonistic robot

Clustering and dimensionality reduction get us closer to human learning, but there's still something very important missing. Children don't just passively observe the world; they do things. They pick up objects they see, play with them, run around, eat, cry, and ask questions. Even the most advanced visual system is of no use to Robby if it doesn't help him interact with the environment. Robby needs to know not just what's where but what to do at each moment. In principle we could teach him using step-by-step instructions, pairing sensor readings with the appropriate actions to take in response, but this is viable only for narrow tasks. The actions you take depend on your goals, not just whatever you are currently perceiving, and those goals can be far in the future. Step-by-step supervision shouldn't be needed, in any case. Parents don't teach their children to crawl, walk, or run; they figure it out on their own. But none of the learning algorithms we've seen so far can do this.

Humans do have one constant guide: their emotions. We seek pleasure and avoid pain. When you touch a hot stove, you instinctively recoil. That's the easy part. The hard part is learning not to touch the stove in the first place. That requires moving to avoid a sharp pain that you have not yet felt. Your brain does this by associating the pain not just with the moment you touch the stove, but with the actions leading up to it. Edward Thorndike called this the law of effect: actions that lead to pleasure are more likely to be repeated in the future; actions that lead to pain, less so. Pleasure travels back through time, so to speak, and actions can eventually become associated with effects that are quite remote from them. Humans can do this kind of long-range reward seeking better than any other animal, and it's crucial to our success. In a famous experiment, children were presented with a marshmallow and told that if they resisted eating it for a few minutes, they could have two. The ones who succeeded went on to do better in school and adult life. Perhaps less obviously, companies using machine learning to improve their websites or their business practices face a similar problem. A company may make a change that brings in more revenue in the short term—like selling an inferior product that costs less to make for the same price as the original superior product—but miss seeing that doing this will lose customers in the longer term.

The learners we saw in the previous chapters are all guided by instant gratification: every action, whether it's flagging a spam e-mail or buying a stock, gets an immediate reward or punishment from the teacher. But there's a whole subfield of machine learning dedicated to algorithms that explore on their own, flail, hit on rewards, and figure out how to get them again in the future, much like babies crawling around and putting things in their mouths.

It's called reinforcement learning, and your first housebot will probably use it a lot. If you ask Robby to make eggs and bacon for you right after you've unpacked him and turned him on, it may take a while. But then, while you're at work, he will explore the kitchen, noting where various things are and what kind of stove you have. By the time you get back, dinner will be ready.

An important precursor of reinforcement learning was a checkers-playing program created by Arthur Samuel, an IBM researcher, in the 1950s. Board games are a great example of a reinforcement learning problem: you have to make a long series of moves without any feedback, and the whole reward or punishment comes at the very end, in the form of a win or loss. Yet Samuel's program was able to teach itself to play as well as most humans. It did not directly learn which move to make in each board position because that would have been too difficult. Rather, it learned how to evaluate each board position—how likely am I to win starting from this position?—and chose the move that led to the best position. Initially, the only positions it knew how to evaluate were the final ones: a win, a tie, or a loss. But once it knew that a certain position was a win, it also knew that positions from which it could move to it were good, and so on. Thomas J. Watson Sr., IBM's president, predicted that when the program was demonstrated IBM stock would go up by fifteen points. It did. The lesson was not lost on IBM, which went on to build a chess champion and a *Jeopardy!* one.

The notion that not all states have rewards (positive or negative) but every state has a value is central to reinforcement learning. In board games, only final positions have a reward (1, 0, or −1 for a win, tie, or loss, say). Other positions give no immediate reward, but they have value in that they can lead to rewards later. A chess position from which you can force checkmate in some number of moves is practically as good as a win and therefore has high value. We can propagate this kind of reasoning all the way to good and bad opening moves, even if at that distance the connection is far from obvious. In video games, the rewards are usually points, and the value of a state is the number of points you can accumulate starting from that state. In real life, a reward now is better than a reward later, so future rewards can be discounted by some rate of return, like investments. Of course, the rewards depend on what actions you choose, and the goal of reinforcement learning is to always choose the action that leads to the greatest rewards. Should you pick up the phone and ask your friend for a date? It could be the start of a beautiful relationship or just the route to a painful rejection. Even if your friend agrees to go on a date, that date may turn out well or not. Somehow, you have to abstract over all the infinite paths the future could take and make a decision now. Reinforcement learning does that by estimating the value of each state—the sum total of the rewards you can expect to get starting from that state—and choosing the actions that maximize it.

Suppose you're moving along a tunnel, Indiana Jones–like, and you come to a fork. Your map says the left tunnel leads to a treasure and the right one to a snake pit. The value of where you're standing—right before the fork—is the value of the treasure because you'll choose to go left. If you always choose the best possible action, then the value of a state differs from the value of the succeeding state only by the immediate reward (if any) that you'll get by performing that action. If we know each state's immediate reward, we can use this observation to update the values of neighboring states, and so on, until all states have consistent values. The treasure's value propagates backward along the tunnel until it reaches the fork and beyond. Once you know the value of each state, you also know which action to choose in each state (the one that maximizes the combination of immediate reward and value of the resulting state). This much was worked out in the 1950s by the control theorist Richard Bellman. But the real problem in reinforcement learning is when you don't have a map of the territory. Then your only choice is to explore and discover what rewards are where. Sometimes you'll discover a treasure, and other times you'll fall into a snake pit. Every time you take an action, you note the immediate reward and the resulting state. That much could be done by supervised learning. But you also update the value of the state you just came from to bring it into line with the value you just observed, namely the reward you got plus the value of the new state you're in. Of course, that value may not yet be the correct one, but if you wander around doing this for long enough, you'll eventually settle on the right values for all the states and the corresponding actions. That's reinforcement learning in a nutshell.

Notice how reinforcement learners face the same exploration-exploitation dilemma we met in Chapter 5: to maximize your rewards, you'll naturally want to always pick the action leading to the highest-value state, but that prevents you from potentially discovering even higher rewards elsewhere. Reinforcement learners solve this by sometimes choosing the best action and sometimes a random one. (The brain even seems to have a "noise generator" for this purpose.) Early on, when there's much to learn, it makes sense to explore a lot. Once you know the territory, it's best to concentrate on exploiting it. That's what humans do over their lifetimes: children explore, and adults exploit (except for scientists, who are eternal children). Children's play is a lot more serious than it looks; if evolution made a creature that is helpless and a heavy burden on its parents for the first several years of its life, that extravagant cost must be for the sake of an even bigger benefit. In effect, reinforcement learning is a kind of speeded-up evolution— trying, discarding, and refining actions within a single lifetime instead of over generations—and by that standard it's extremely efficient.

Research on reinforcement learning started in earnest in the early 1980s, with the work of Rich Sutton and Andy Barto at the University of Massachusetts. They felt that learning depends crucially on interacting with the environment, but supervised algorithms didn't capture this, and they found inspiration instead in the psychology of animal learning. Sutton went on to become the leading proponent of reinforcement learning. Another key step happened in 1989, when Chris Watkins at Cambridge, initially motivated by his experimental observations of children's learning, arrived at the modern formulation of reinforcement learning as optimal control in an unknown environment.

Reinforcement learners as we've seen them so far are not very realistic, however, because they don't know what to do in a state unless they've been there before, and in the real world no two situations are ever exactly alike. We need to be able to generalize from previously visited states to new ones. Luckily, we already know how to do that: all we have to do is wrap reinforcement learning around one of the supervised learners we've met before, such as a multilayer perceptron. The neural network's job is now to predict the value of a state, and the error signal for backpropagation is the difference between the predicted and observed values. There's a problem, however. In supervised learning the target value for a state is always the same, but in reinforcement learning, it keeps changing as a consequence of updates to nearby states. As a result, reinforcement learning with generalization often fails to settle on a stable solution, unless the inner learner is something very simple, like a linear function. Nevertheless, reinforcement learning with neural networks has had some notable successes. An early one was a human-level backgammon player. More recently, a reinforcement learner from DeepMind, a London-based startup, beat an expert human player at Pong and other simple arcade games. It used a deep network to predict actions' values from the console screen's raw pixels. With its end-to-end vision, learning, and control, the system bore at least a passing resemblance to an artificial brain. This may help explain why Google paid half a billion dollars for DeepMind, a company with no products, no revenues, and few employees.

Gaming aside, researchers have used reinforcement learning to balance poles, control stick-figure gymnasts, park cars backward, fly helicopters upside down, manage automated telephone dialogues, assign channels in cell phone networks, dispatch elevators, schedule space-shuttle cargo loading, and much else. Reinforcement learning has also influenced psychology and neuroscience. The brain does it, using the neurotransmitter dopamine to propagate differences between expected and actual rewards. Reinforcement learning explains Pavlovian conditioning, but unlike behaviorism, it allows animals to have internal mental states. Foraging bees use it, as do mice finding cheese in mazes. Your daily life is a stream of little-noticed miracles made possible in part by reinforcement learning. You get up, get dressed, eat breakfast, and drive to work, all the while thinking about something else. Below the surface, reinforcement learning continually orchestrates and fine-tunes this prodigious symphony of motion. Snippets of reinforcement learning, also known as habits, make up most of what you do. You feel hungry, walk to the fridge, and grab a snack. As Charles Duhigg shows in *The Power of Habit*, understanding and controlling this cycle of cue, routine, and reward is key to success, not just for individuals but for businesses and even whole societies.

Of reinforcement learning's founders, Rich Sutton is the most gung ho. For him, reinforcement learning is the Master Algorithm and solving it is tantamount to solving AI. Chris Watkins, on the other hand, is dissatisfied. He sees many things children can do that reinforcement learners can't: solve problems, solve them better after a few attempts, make plans, acquire increasingly abstract knowledge. Luckily, we also have learning algorithms for these higher-level abilities, the most important of which is chunking.

Practice makes perfect

To learn is to get better with practice. You may barely remember it now, but learning to tie your shoelaces was really hard. At first you couldn't do it at all, despite your five years of age. Then your laces probably came undone faster than you could tie them. But little by little you learned to tie them faster and better until it became completely automatic. The same happens with lots of other things, like crawling, walking, running, riding a bike, and driving a car; reading, writing, and arithmetic; playing an instrument and practicing a sport; cooking and using a computer. Ironically, you learn the most when it's most painful: early on, when every step is difficult, you keep failing, and even when you succeed, the results are not very pretty. After you've mastered your golf swing or tennis serve, you can spend years perfecting it, but all those years make less difference than the first few weeks did. You get better with practice, but not at a constant rate: at first you improve quickly, then not so quickly, then very slowly. Whether it's playing games or the guitar, the curve of performance improvement over time—how well you do something or how long it takes you to do it—has a very specific form:

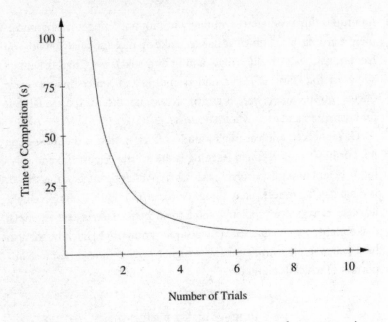

This type of curve is called a power law, because performance varies as time raised to some negative power. For example, in the figure above, time to completion is proportional to the number of trials raised to minus two (or equivalently, one over the number of trials squared). Pretty much every human skill follows a power law, with different powers for different skills. (In contrast, Windows never gets faster with practice—something for Microsoft to work on.)

In 1979, Allen Newell and Paul Rosenbloom started wondering what could be the reason for this so-called power law of practice. Newell was one of the founders of AI and a leading cognitive psychologist, and Rosenbloom was one of his graduate students at Carnegie Mellon University. At the time, none of the existing models of practice could explain the power law. Newell and Rosenbloom suspected it might have something to do with chunking, a concept from the psychology of perception and memory. We perceive and remember things in chunks, and we can only hold so many chunks in short-term memory at any given time (seven plus or minus two, according to the classic paper by George Miller). Crucially, grouping things into chunks allows us to process much more information than we otherwise could. That's why telephone numbers have hyphens: 1-723-458-3897 is much easier to remember than 17234583897. Herbert Simon, Newell's longtime collaborator and AI cofounder, had earlier found that the main difference between novice and expert chess players is that novices perceive chess positions one piece at a time while experts see larger patterns involving multiple pieces. Getting better at chess mainly involves acquiring more and larger such chunks. Newell and Rosenbloom hypothesized that a similar process is at work in all skill acquisition, not just chess.

In perception and memory, a chunk is just a symbol that stands for a pattern of other symbols, like AI stands for artificial intelligence. Newell and Rosenbloom adapted this notion to the theory of problem solving that Newell and Simon had developed earlier. Newell and Simon asked experimental subjects to solve problems—for example, derive one mathematical formula from another on the blackboard—while narrating aloud how they were going about it. They found that humans solve problems by decomposing them into subproblems, subsubproblems, and so on and systematically reducing the differences between the initial state (the first formula, say) and the goal state (the second formula). Doing so requires searching for a sequence of actions that will work, however, and that takes time. Newell and Rosenbloom's hypothesis was that each time we solve a subproblem, we form a chunk that allows us to go directly from the state before we solve it to the state after. A chunk in this sense has two parts: the stimulus (a pattern you recognize in the external world or in your short-term memory) and the response (the sequence of actions you execute as a result). Once you've learned a chunk, you store it in long-term memory. Next time you have to solve the same subproblem, you can just apply the chunk, and save the time spent searching. This happens at all levels until you have a chunk for the whole problem and can solve it automatically. To tie your shoelaces, you tie the starting knot, make a loop with one end, wrap the other end around it, and pull it through the hole in the middle. Each of these is far from trivial for a five-year-old, but once you've acquired the corresponding chunks, you're almost there.

Rosenbloom and Newell set their chunking program to work on a series of problems, measured the time it took in each trial, and lo and behold, out popped a series of power law curves. But that was only the beginning. Next they incorporated chunking into Soar, a general theory of cognition that Newell had been working on with John Laird, another one of his students. Instead of working only within a predefined hierarchy of goals, the Soar program could define and solve a new subproblem every time it hit a snag. Once it formed a new chunk, Soar generalized it to apply to similar problems, in a manner similar to inverse deduction. Chunking in Soar turned out to be a good model of lots of learning phenomena besides the power law of practice. It could even be applied to learning new knowledge by chunking data and analogies. This led Newell, Rosenbloom, and Laird to hypothesize that chunking is the *only* mechanism needed for learning—in other words, the Master Algorithm.

Being classic AI types, Newell, Simon, and their students and followers were strong believers in the primacy of problem solving. If the problem solver is powerful, the learner can piggyback on it and be simple. Indeed, learning is just another kind of problem solving. Newell and company made a concerted effort to reduce all learning to chunking and all cognition to Soar, but in the end they failed. One problem was that, as the problem solver learned more chunks, and more complicated ones, the cost of trying them often became so high that the program got slower instead of faster. Somehow humans avoid this, but so far researchers in this area have not figured out how. On top of that, trying to reduce reinforcement learning, supervised learning, and everything else to chunking ultimately created more problems than it solved. Eventually, the Soar researchers conceded defeat and incorporated those other types of learning into Soar as separate mechanisms. Nevertheless, chunking remains a preeminent example of a learning algorithm inspired by psychology, and the true Master Algorithm, whatever it turns out to be, must surely share its ability to improve with practice.

Chunking and reinforcement learning are not as widely used in business as supervised learning, clustering, or dimensionality reduction, but a simpler type of learning by interacting with the environment is: learning the effects of your actions (and acting accordingly). If the background color of your e-commerce site's home page is currently blue and you're wondering whether making it red would increase sales, try it out on a hundred thousand randomly chosen customers and compare the results with those of the regular site. This technique, called A/B testing, was at first used mainly in drug trials but has since spread to many fields where data can be gathered on demand, from marketing to foreign aid. It can also be generalized to try many combinations of changes at once, without losing track of which changes lead to which gains (or losses). Companies like Amazon and Google swear by it; you've probably participated in thousands of A/B tests without realizing it. A/B testing gives the lie to the oft-heard criticism that big data is only good for finding correlations, not causation. Philosophical fine points aside, learning causality is learning the effects of your actions, and anyone with a stream of data they can affect can do it—from a one-year-old splashing around in the bathtub to a president campaigning for reelection.

Learning to relate

If we endow Robby the robot with all the learning abilities we've seen so far in this book, he'll be pretty smart but still a bit autistic. He'll see the world as a bunch of separate objects, which he can identify, manipulate, and even make predictions about, but he won't understand that the world is a web of interconnections. Robby the doctor would be very good at diagnosing someone with the flu based on his symptoms but unable to suspect that the patient has swine flu because he has been in contact with someone infected with it. Before Google, search engines decided whether a web page was relevant to your query by looking at its content—what else? Brin and Page's insight was that the strongest sign a page is relevant is that relevant pages link to it. Similarly, if you want to predict whether a teenager is at risk of starting to smoke, by far the best thing you can do is check whether her close friends smoke. An enzyme's shape is as inseparable from the shapes of the molecules it brings together as a lock is from its key. Predator and prey have deeply entwined properties, each evolved to defeat the other's properties. In all of these cases, the best way to understand an entity—whether it's a person, an animal, a web page, or a molecule—is to understand how it relates to other entities. This requires a new kind of learning that doesn't treat the data as a random sample of unrelated objects but as a glimpse into a complex network. Nodes in the network interact; what you do to one affects the others and comes back to affect you. Relational learners, as they're called, may not quite have social intelligence, but they're the next best thing. In traditional statistical learning, every man is an island, entire of itself. In relational learning, every man is a piece of the continent, a part of the main. Humans are relational learners, wired to connect, and if we want Robby to grow into a perceptive, socially adept robot, we need to wire him to connect, too.

The first difficulty we face is that, when the data is all one big network, we no longer seem to have many examples to learn from, just one—and that's not enough. Naïve Bayes learns that a fever is a symptom of the flu by counting the number of fever-stricken flu patients. If it could only see one patient, it would either conclude that flu always causes fever or that it never does, both of which are wrong. We would like to learn that the flu is contagious by looking at the pattern of infections in a social network—a clump of infected people here, a clump of uninfected ones there—but we only have one pattern to look at, even if it's in a network of seven billion people, so it's not clear how to generalize. The key is to notice that, embedded in that big network, we have many examples of *pairs* of people. If acquaintances are more likely to both have the flu than pairs of people who have never met, then being acquainted with a flu patient makes you more likely to be one as well. Unfortunately, however, we can't just count how many pairs of acquaintances in the data both have the flu and turn those counts into probabilities. This is because a person has many acquaintances, and all the pairwise probabilities don't add up to a coherent model that lets us, for example, compute how likely someone is to have the flu given which of their acquaintances do. We didn't have this problem when the examples were all separate, and we wouldn't have it in, say, a society of childless couples, each living on their own desert island. But that's not the real world, and there wouldn't be any epidemics in it, anyway.

The solution is to have a set of features and learn their weights, as in Markov networks. For every person X, we can have the feature *X has the flu*; for every pair of acquaintances X and Y, the feature *X and Y both have the flu*; and so on. As in Markov networks, the maximum-likelihood weights are the ones that make each feature occur with the frequency observed in the data. The weight of *X has the flu* will be high if a lot of people have the flu. The weight of *X and Y both have the flu* will be high if, when person X has the flu, the odds that acquaintance Y also has the flu are higher than for a randomly chosen member of the network. If 40 percent of people have the flu and so do 16 percent of all acquaintance pairs, then the weight of *X and Y both have the flu* will be zero, because we don't need that feature to correctly reproduce the data's statistics ($0.4 \times 0.4 = 0.16$). But if the feature has a positive weight, flu is more likely to occur in clumps than to just infect people at random, and you're more likely to have the flu if your acquaintances do.

Notice that the network has a separate feature for each pair of people: *Alice and Bob both have the flu, Alice and Chris both have the flu,* and so on. But we can't learn a separate weight for each pair, because we only have one data point per pair (whether it's infected or not), and we wouldn't be able to generalize to members of the network we haven't diagnosed yet (do Yvette and Zach both have the flu?). What we can do instead is learn a single weight for all features of the same form, based on all the instances of it that we've seen. In effect, *X and Y have the flu* is a template for features that can be instantiated with each pair of acquaintances (Alice and Bob, Alice and Chris, etc.). The weights for all the instances of a template are "tied together," in the sense that they all have the same value, and that's how we can generalize despite having only one example (the whole network). In nonrelational learning, the parameters of a model are tied in only one way: across all the independent examples (e.g., all the patients we've diagnosed). In relational learning, every feature template we create ties the parameters of all its instances.

We're not limited to pairwise or individual features. Facebook wants to predict who your friends are so it can recommend them to you. It can use the rule *Friends of friends are likely to be friends* for that, but each instance of it involves three people: if Alice and Bob are friends, and Bob and Chris are also friends, then Alice and Chris are potential friends. H. L. Mencken's quip that a man is wealthy if he makes more than his wife's sister's husband involves four people. Each of these rules can be turned into a feature template in a relational model, and a weight for it can be learned based on how often the feature occurs in the data. As in Markov networks, the features themselves can also be learned from the data.

Relational learners can generalize from one network to another (e.g., learn a model of how flu spreads in Atlanta and apply it in Boston). They can also learn on more than one network (e.g., Atlanta and Boston, assuming, unrealistically, that no one in Atlanta is ever in contact with anyone in Boston). But unlike "regular" learning, where all examples must have exactly the same number of attributes, in relational learning networks can vary in size; a larger network will just have more instances of the same templates than a smaller one. Of course, the generalization from a smaller network to a larger one may or may not be accurate, but the point is that nothing prevents it; and large networks often do behave locally like small ones.

The neatest trick a relational learner can do is to turn a sporadic teacher into an assiduous one. For an ordinary classifier, examples without classes are useless. If I'm given a patient's symptoms, but not the diagnosis, that doesn't help me learn to diagnose. But if I know that some of the patient's friends have the flu, that's indirect evidence that he may have the flu as well. Diagnosing a few people in a network and then propagating those diagnoses to their friends, and their friends' friends, is the next best thing to diagnosing everyone. The inferred diagnoses may be noisy, but the overall statistics of how symptoms correlate with the flu will probably be a lot more accurate and complete than if I had only a handful of isolated diagnoses to draw on. Children are very good at making the most of the sporadic supervision they get (provided they don't choose to ignore it). Relational learners share some of that ability.

All this power comes at a cost, however. In an ordinary classifier, such as a decision tree or a perceptron, inferring an entity's class from its attributes is a matter of a few lookups and a bit of arithmetic. In a network, each node's class depends indirectly on all the others', and we can't infer it in isolation. We can resort to the same kinds of inference techniques we used for Bayesian networks, like loopy belief propagation or MCMC, but the scale is different. A typical Bayesian network has perhaps thousands of variables, but a typical social network has millions of nodes or more. Luckily, because the model of the network consists of many repetitions of the same features with the same weights, we can often condense the network into "supernodes," each consisting of many nodes that we know will have the same probabilities, and solve a much smaller problem with the same result.

Relational learning has a long history, going back to at least the seventies and symbolist techniques like inverse deduction. But it acquired a new impetus with the advent of the Internet. Suddenly networks were everywhere, and modeling them was urgent. One phenomenon I found particularly intriguing was word of mouth. How does information propagate in a social network? Can we measure each member's influence and target just enough of the most influential members to set off a wave of word of mouth? With my student Matt Richardson, I designed an algorithm that did just that. We applied it to Epinions, a product review site that allowed members to say whose reviews they trusted. We found, among other things, that marketing a product to the single most influential member—trusted by many followers who were in turn trusted by many others, and so on—was as good as marketing to a third of all the members in isolation. An avalanche of other research on this problem followed. Since then, I've applied relational learning to many others, including predicting who will form links in a social network, integrating databases, and enabling robots to build maps of their surroundings.

If you want to understand how the world works, relational learning is a good tool to have. In Isaac Asimov's *Foundation*, the scientist Hari Seldon manages to mathematically predict the future of humanity and thereby save it from decadence. Paul Krugman, among others, has confessed that this seductive dream was what made him become an economist. According to Seldon, people are like molecules in a gas, and the law of large numbers ensures that even if individuals are unpredictable, whole societies aren't. Relational learning reveals why this is not the case. If people were independent, each making decisions in isolation, societies would indeed be predictable, because all those random decisions would add up to a fairly constant average. But when people interact, larger assemblies can be less predictable than smaller ones, not more. If confidence and fear are contagious, each will dominate for a while, but every now and then an entire society will swing from one to the other. It's not all bad news, though. If we can measure how strongly people influence each other, we can estimate how long it will be before a swing occurs, even if it's the first one—another way in which black swans are not necessarily unpredictable.

A common complaint about big data is that the more data you have, the easier it is to find spurious patterns in it. This may be true if the data is just a huge set of disconnected entities, but if they're interrelated, the picture changes. For example, critics of using data mining to catch terrorists argue that, ethical issues aside, it will never work because there are too many innocents and too few terrorists and so mining for suspicious patterns will either cause too many false alarms or never catch anyone. Is someone videotaping the New York City Hall a tourist or a terrorist scoping out a bombing site? And is someone buying large quantities of ammonium nitrate a farmer or a bomb maker? Each of these looks innocent enough in isolation, but if the "tourist" and the "farmer" have been in close phone contact, and the latter just drove his heavily laden pickup into Manhattan, maybe it's time for someone to take a closer look. The NSA likes to mine records of who called whom not just because it's arguably legal, but because they're often more informative to the prediction algorithms than the content of the calls, which it would take a human to understand.

Social networks aside, the killer app of relational learning is understanding how living cells work. A cell is a complex metabolic network with genes coding for proteins that regulate other genes, long interlocking chains of chemical reactions, and products migrating from one organelle to another. Independent entities, doing their work in isolation, are nowhere to be seen. A cancer drug must disrupt cancer cells' workings without interfering with normal ones'. If we have an accurate relational model of both, we can try many different drugs *in silico*, letting the model infer their good and bad effects and keeping only the best ones to try *in vitro* and finally *in vivo*.

Like human memory, relational learning weaves a rich web of associations. It connects percepts, which a robot like Robby can acquire by clustering and dimensionality reduction, with skills, which he can learn by reinforcement and chunking, and with the higher-level knowledge that comes from reading, going to school, and interacting with humans. Relational learning is the last piece of the puzzle, the final ingredient we need for our alchemy. And now it's time to repair to the lab and transmute all these elements into the Master Algorithm.

CHAPTER NINE

The Pieces of the Puzzle Fall into Place

Machine learning is both a science and a technology, and both characteristics give us hints on how to unify it. On the science side, unifying theories often begin with a deceptively simple observation. Two seemingly unrelated phenomena turn out to be just two faces of the same coin, and like the first domino to fall, that realization sets off a cascade of others. An apple falling to the ground, the moon hanging in the sky: both are caused by gravity, and—apocryphal story or not—once Newton figured out how, gravity turned out to also account for the tides, the precession of the equinoxes, the trajectories of comets, and much else. In everyday experience, electricity and magnetism are never seen together: a lightning spark here, a rock that attracts iron objects there, both quite rare. But once Maxwell figured out how a changing electric field gives rise to magnetism and vice versa, it became clear that light itself is an intimate marriage of the two, and today we know that, far from rare, electromagnetism pervades all matter. Mendeleev's periodic table not only organized all the known elements into just two dimensions, it also predicted where new elements would be found. Darwin's observations aboard the *Beagle* suddenly began to make sense when Malthus's *Essay on Population* suggested natural selection as the organizing principle.

When Crick and Watson hit on the double helix structure as an explanation for the puzzling properties of DNA, they immediately saw how it might replicate itself, and biology's transition from stamp collecting (in Rutherford's pejorative words) to unified science had begun. In each of these cases, a bewildering variety of observations turned out to have a common cause, and once scientists identified it, they could in turn use it to predict many new phenomena. Similarly, even though the learners we've met in this book seem quite disparate—some based on the brain, some on evolution, some on abstract mathematical principles—they in fact have much in common, and the resulting theory of learning yields many new insights.

Although it is less well known, many of the most important technologies in the world are the result of inventing a unifier, a single mechanism that does what previously required many. The Internet, as the name implies, is a network that interconnects networks. Without it, every type of network would need a different protocol to talk to every other, much like we need a different dictionary for every pair of languages in the world. The Internet's protocols are an Esperanto that gives each computer the illusion of talking directly to any other and that allows e-mail and the web to ignore the details of the physical infrastructure they flow over. Relational databases do something similar for enterprise applications, allowing developers and users to think in terms of the abstract relational model and ignore the different ways computers go about answering queries. A microprocessor is an assembly of digital electronic components that can mimic any other assembly. Virtual machines allow the same computer to pose as a hundred different computers to a hundred different people at the same time, and help make the cloud possible. Graphical user interfaces let us edit documents, spreadsheets, slide decks, and much else using a common language of windows, menus, and mouse clicks. The computer itself is a unifier: a single device capable of solving any logical or mathematical problem, provided we know how to program it. Even plain old electricity is a kind of unifier: you can generate it from many different sources—coal, gas, nuclear, hydro, wind, solar—and consume it in an infinite variety of ways.

A power station doesn't know or care how the electricity it produces will be consumed, and your porch light, dishwasher, or brand-new Tesla are oblivious to where their electricity supply comes from. Electricity is the Esperanto of energy. The Master Algorithm is the unifier of machine learning: it lets any application use any learner, by abstracting the learners into a common form that is all the applications need to know.

Our first step toward the Master Algorithm will be surprisingly simple. As it turns out, it's not hard to combine many different learners into one, using what is known as metalearning. Netflix, Watson, Kinect, and countless others use it, and it's one of the most powerful arrows in the machine learner's quiver. It's also a stepping-stone to the deeper unification that will follow.

Out of many models, one

Here's a challenge: you have fifteen minutes to combine decision trees, multilayer perceptrons, classifier systems, Naïve Bayes, and SVMs into a single algorithm possessing the best properties of each. Quick—what can you do? Clearly, it can't involve the details of the individual algorithms; there's no time for that. But how about the following? Think of each learner as an expert on a committee. Each looks carefully at the instance to be classified—what is the diagnosis for this patient?—and confidently makes its prediction. You're not an expert yourself, but you're the chair of the committee, and your job is to combine their recommendations into a final decision. What you have on your hands is in fact a new classification problem, where instead of the patient's symptoms, the input is the experts' opinions. But you can apply machine learning to this problem in the same way the experts applied it to the original one. We call this metalearning because it's learning about the learners. The metalearner can itself be any learner, from a decision tree to a simple weighted vote. To learn the weights, or the decision tree, we replace the attributes of each original example by the learners' predictions. Learners that often predict the correct class will get high weights, and inaccurate ones will tend to be ignored. With a decision tree, the choice of whether to use a learner can be contingent on other learners' predictions. Either way, to obtain a learner's prediction for a given training example, we must first apply it to the original training set *excluding that example* and use the resulting classifier—otherwise the committee risks being dominated by learners that overfit, since they can predict the correct class just by remembering it. The Netflix Prize winner used metalearning to combine hundreds of different learners. Watson uses it to choose its final answer from the available candidates. Nate Silver combines polls in a similar way to predict election results.

This type of metalearning is called stacking and is the brainchild of David Wolpert, whom we met in Chapter 3 as the author of the "no free lunch" theorem. An even simpler metalearner is bagging, invented by the statistician Leo Breiman. Bagging generates random variations of the training set by resampling, applies the same learner to each one, and combines the results by voting. The reason to do this is that it reduces variance: the combined model is much less sensitive to the vagaries of the data than any single one, making this a remarkably easy way to improve accuracy. If the models are decision trees and we further vary them by withholding a random subset of the attributes from consideration at each node, the result is a so-called random forest. Random forests are some of the most accurate classifiers around. Microsoft's Kinect uses them to figure out what you're doing, and they regularly win machine-learning competitions.

One of the cleverest metalearners is boosting, created by two learning theorists, Yoav Freund and Rob Schapire. Instead of combining different learners, boosting repeatedly applies the same classifier to the data, using each new model to correct the previous ones' mistakes. It does this by assigning weights to the training examples; the weight of each misclassified example is increased after each round of learning, causing later rounds to focus more on it. The name *boosting* comes from the notion that this process can boost a classifier that's only slightly better than random guessing, but consistently so, into one that's almost perfect.

Metalearning is remarkably successful, but it's not a very deep way to combine models. It's also expensive, requiring as it does many runs of learning, and the combined models can be quite opaque. ("I believe you have prostate cancer because the decision tree, the genetic algorithm, and Naïve Bayes say so, although the multilayer perceptron and the SVM disagree.") Moreover, all the combined models are really just one big, messy model. Can't we have a single learner that does the same job? Yes we can.

The Master Algorithm

Our unified learner is perhaps best introduced through an extended allegory. If machine learning is a continent divided into the territories of the five tribes, the Master Algorithm is its capital city, standing on the unique spot where the five territories meet. As you approach it from a distance, you can see that the city is made up of three concentric circles, each bounded by a wall. The outer and by far widest circle is Optimization Town. Each house here is an algorithm, and they come in all shapes and sizes. Some are under construction, the locals busy around them; some are gleaming new; and some look old and abandoned. Higher up the hill lies the Citadel of Evaluation. From its mansions and palaces orders issue continuously to the algorithms below. Above all, silhouetted against the sky, rise the Towers of Representation. Here live the rulers of the city. Their immutable laws set forth what can and cannot be done not just in the city but throughout the continent. Atop the central, tallest tower flies the flag of the Master Algorithm, red and black, with a five-pointed star surrounding an inscription that you cannot yet make out.

The city is divided into five sectors, each belonging to one of the five tribes. Each sector stretches down from its Tower of Representation to the city's outer walls, encompassing the tower, a clutch of palaces in the Citadel of Evaluation, and the streets and houses in Optimization Town they overlook. The five sectors and three rings divide the city into fifteen districts, fifteen shapes, fifteen pieces of the puzzle you need to solve:

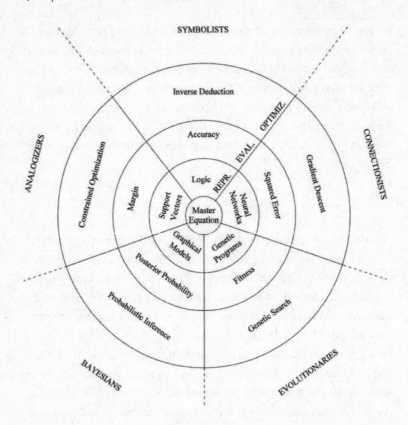

You gaze intently at the map, trying to decipher its secret. The fifteen pieces all match quite precisely, but you need to figure out how they combine to form just three: the representation, evaluation, and optimization components of the Master Algorithm. Every learner has these three elements, but they vary from tribe to tribe.

Representation is the formal language in which the learner expresses its models. The symbolists' formal language is logic, of which rules and decision trees are special cases. The connectionists' is neural networks. The evolutionaries' is genetic programs, including classifier systems. The Bayesians' is graphical models, an umbrella term for Bayesian networks and Markov networks. The analogizers' is specific instances, possibly with weights, as in an SVM.

The evaluation component is a scoring function that says how good a model is. Symbolists use accuracy or information gain. Connectionists use a continuous error measure, such as squared error, which is the sum of the squares of the differences between the predicted values and the true ones. Bayesians use the posterior probability. Analogizers (at least of the SVM stripe) use the margin. In addition to how well the model fits the data, all tribes take into account other desirable properties, such as the model's simplicity.

Optimization is the algorithm that searches for the highest-scoring model and returns it. The symbolists' characteristic search algorithm is inverse deduction. The connectionists' is gradient descent. The evolutionaries' is genetic search, including crossover and mutation. The Bayesians are unusual in this regard: they don't just look for the best model, but average over all models, weighted by how probable they are. To do the weighting efficiently, they use probabilistic inference algorithms like MCMC. The analogizers (or more precisely, the SVM mavens) use constrained optimization to find the best model.

After a long day's journey, the sun is rapidly nearing the horizon, and you need to hurry before it gets dark. The city's outer wall has five massive gates, each controlled by one of the tribes and leading to its district in Optimization Town. Let us enter through the Gradient Descent Gate, after whispering the watchword—"deep learning"—to the guard, and spiral in toward the Towers of Representation. From the gate the street ascends steeply up the hill to the citadel's Squared Error Gate, but instead you turn left toward the evolutionary sector. The houses in the gradient descent district are all smooth curves and densely intertwined patterns, almost more like a jungle than a city. But when gradient descent gives way to genetic search, the picture changes abruptly. Here the houses rise higher, with structure piled on structure, but the structures are spare, almost vacant, as if waiting to be filled in by gradient descent's curves. That's it: the way to combine the two is to use genetic search to find the structure of the model and let gradient descent fill in its parameters. This is what nature does: evolution creates brain structures, and individual experience modulates them.

The first step accomplished, you hurry on to the Bayesian district. Even from a distance, you can see how it clusters around the Cathedral of Bayes' Theorem. MCMC Alley zigzags randomly along the way. This is going to take a while. You take a shortcut onto Belief Propagation Street, but it seems to loop around forever. Then you see it: the Most Likely Avenue, rising majestically toward the Posterior Probability Gate. Rather than average over all models, you can head straight for the most probable one, confident that the resulting predictions will be almost the same. And you can let genetic search pick the model's structure and gradient descent its parameters. With a sigh of relief, you realize that's all the probabilistic inference you'll need, at least until it's time to answer questions using the model.

You keep going. The constrained optimization district is a maze of narrow alleys and dead ends, examples of all kinds standing cheek by jowl everywhere, with an occasional clearing around a support vector. Clearly, all you need to do to avoid bumping into examples of the wrong class is add constraints to the optimizer you've already assembled. But come to think of it, not even that is necessary. When we learn SVMs, we usually let margins be violated in order to avoid overfitting, provided each violation pays a penalty. In this case the optimal example weights can again be learned by a form of gradient descent. That was easy. You feel like you're starting to get the hang of it.

The dense ranks of instances end abruptly, and you find yourself in the inverse deduction district, a place of broad avenues and ancient stone buildings. The architecture here is geometric, austere, made of straight lines and right angles. Even the severely pruned trees have rectangular trunks, and their leaves are meticulously labeled with class predictions. The denizens of this district seem to build their houses in a peculiar way: they start with the roof, which they label "Conclusions," and gradually fill in the gaps between it and the ground, which they label "Premises." One by one, they find a stone block that's the right shape to fill in a particular gap and hoist it up to its place. But, you notice, many gaps have the same shape, and it would be faster to cut and combine blocks until they form that shape, and then repeat the process as many times as necessary. In other words, you could use genetic search to do inverse deduction. Neat. It looks like you've boiled down the five optimizers to a simple recipe: genetic search for structure and gradient descent for parameters. And even that may be overkill. For a lot of problems, you can whittle genetic search down to hill climbing if you do three things: leave out crossover, try all possible point mutations in each generation, and always select the single best hypothesis to seed the next generation.

What's that statue up ahead? Aristotle, looking rather disapprovingly toward the tangled mess of the gradient descent quarter. You've come full circle. You have the unified optimizer you need for the Master Algorithm, but this is no time to congratulate yourself. Night has fallen, and you still have much to do. You enter the Citadel of Evaluation through the imposing but rather narrow Accuracy Gate. The inscription above it says "Abandon all hope of overfitting, ye who enter here." As you circle past the palaces of the five tribes' evaluators, you mentally snap the pieces into place. You use accuracy to evaluate yes-or-no predictions and squared error for continuous ones. Fitness is just the evolutionaries' name for the scoring function; you can make it anything you want, including accuracy and squared error. Posterior probability reduces to squared error if you ignore the prior probability and the errors follow a normal distribution. The margin, if you allow it to be violated for a price, becomes a softer version of accuracy: instead of paying no penalty for a correct prediction and a penalty of one for an incorrect prediction, the penalty is zero until you get inside the margin, at which point it starts to steadily go up. Whew! Combining the evaluators was a lot easier than combining the optimizers. But the Towers of Representation, looming above you, fill you with a sense of foreboding.

You've reached the final stage of your quest. You knock on the door of the Tower of Support Vectors. A menacing-looking guard opens it, and you suddenly realize that you don't know the password. "Kernel," you blurt out, trying to keep the panic from your voice. The guard bows and steps aside. Regaining your composure, you step in, mentally kicking yourself for your carelessness. The entire ground floor of the tower is taken up by a lavishly appointed circular chamber, with what seems to be a marble representation of an SVM occupying pride of place at the center. As you walk around it, you notice a door on the far side. It must lead to the central tower—the Tower of the Master Algorithm. The door seems unguarded. You decide to take a shortcut. Slipping through the doorway, you walk down a short corridor and find yourself in an even larger pentagonal chamber, with a door in each wall. In the center, a spiral staircase rises as high as the eye can see. You hear voices above and duck into the doorway opposite. This one leads to the Tower of Neural Networks. Once again you're in a circular chamber, this one with a sculpture of a multilayer perceptron as the centerpiece. Its parts are different from the SVM's, but their arrangement is remarkably similar. Suddenly you see it: an SVM is just a multilayer perceptron with a hidden layer composed of kernels instead of S curves and an output that's a linear combination instead of another S curve.

Could it be that the other representations also have a similar form? With rising excitement, you run back through the pentagonal chamber and into the Tower of Logic. Staring at the depiction of a set of rules in the center, you try to discern a pattern. Yes! Each rule is just a highly stylized neuron. For example, the rule *If it's a giant reptile and breathes fire then it's a dragon* is just a perceptron with weights of one for *it's a giant reptile* and *breathes fire* and a threshold of 1.5. And a set of rules is a multilayer perceptron with a hidden layer containing one neuron for each rule and an output neuron to form the disjunction of the rules. There's a nagging doubt in the back of your mind, but you don't have time for it right now. As you cross the pentagonal chamber to the Tower of Genetic Programs, you can already see how to bring them into the fold. Genetic programs are just programs, and programs are just logic constructs. The sculpture of a genetic program in the chamber is in the shape of a tree, subroutines branching into more subroutines, and when you look closely at the leaves, you can see that they're just simple rules. So programs boil down to rules, and if rules can be reduced to neurons, so can programs.

On to the Tower of Graphical Models. Unfortunately, the sculpture in its circular chamber looks nothing like the others. A graphical model is a product of factors: conditional probabilities, in the case of Bayesian networks, and non-negative functions of the state, in the case of Markov networks. Try as you might, you just can't see the connection to neural networks or sets of rules. Disappointment washes over you. But then you put on your "loggles," which replace every function by its logarithm. Eureka—the product of factors is now a sum of terms, just like an SVM, a voting set of rules, or a multilayer perceptron without the output S curve. For example, you can translate a Naïve Bayes dragon classifier into a perceptron whose weight for *breathes fire* is the log of *P(breathes fire | dragon)* minus the log of *P(breathes fire | not dragon)*. But of course, graphical models are much more general than this because they can represent probability distributions over many variables, not just the distribution of one variable (the class) given the others (the attributes).

You did it! Or did you? Absorbing SVMs into neural networks and neural networks into graphical models: that worked. So did absorbing genetic programs into logic. But combining logic and graphical models? Something is amiss there. Belatedly, you see the problem: logic has a dimension that graphical models lack and vice versa. The sculptures in the five chambers matched because they were simple allegories, but the reality doesn't. Graphical models don't let us represent rules involving more than one object, like *Friends of friends are friends*; all their variables have to be properties of the same object. They also can't represent arbitrary programs, which pass sets of variables from one subroutine to another. Logic can easily do both of these things, but on the other hand it can't represent uncertainty, ambiguity, or degrees of similarity. And without a representation that can do all of these things, you don't have a universal learner.

You rack your brains for a solution, but the more you try, the harder it gets. Perhaps unifying logic and probability is just beyond human ability. Exhausted, you fall asleep. A deep growl jolts you awake. The hydra-headed complexity monster pounces on you, jaws snapping, but you duck at the last moment. Slashing desperately at the monster with the sword of learning, the only one that can slay it, you finally succeed in cutting off all its heads. Before it can grow new ones, you run up the stairs.

After an arduous climb, you reach the top. A wedding is in progress. Praedicatus, First Lord of Logic, ruler of the symbolic realm and Protector of the Programs, says to Markovia, Princess of Probability, Empress of Networks: "Let us unite our realms. To my rules thou shalt add weights, begetting a new representation that will spread far across the land." The princess says, "And we shall call our progeny Markov logic networks."

Your head is spinning. You go outside to the balcony. The sun has risen over the city. You gaze out over the rooftops to the countryside beyond. Forests of servers stretch away in all directions, humming quietly, waiting for the Master Algorithm. Convoys move along the roads, carrying gold from the data mines. Far to the west, the land gives way to a sea of information, dotted with ships. You look up at the flag of the Master Algorithm. You can now clearly see the inscription inside the five-pointed star:

$$P = e^{w \cdot n} / Z$$

What could this mean, you wonder?

Markov logic networks

In 2003, I started thinking about the problem of how to unify logic and probability, together with my student Matt Richardson. At first we made little progress because we were trying to do it with Bayesian networks, and their rigid form—a strict order on variables, conditional distributions of children given parents—is incompatible with the flexibility of logic. But the day before Christmas Eve, I realized there was a much better way. If we switched to Markov networks, we could use *any* logical formula as a template for Markov network features, and that would unify logic and graphical models. Let's see how.

Recall that a Markov network is defined by a weighted sum of features, much like a perceptron. Suppose we have a collection of photos of people. We pick a random one and compute features of it like *The person has gray hair, The person is old, The person is a woman,* and so on. In a perceptron, we pass the weighted sum of these features through a threshold to decide whether, say, the person is your grandmother or not. In a Markov network, we do something very different (at least at first sight): we exponentiate the weighted sum, turning it into a product of factors, and this product is the probability of choosing that particular picture from the collection, regardless of whether your grandmother is in it. If you have many pictures of old people, the weight of that feature goes up. If most of them are of men, the weight of *The person is a woman* goes down. The features can be anything we want, making Markov networks a remarkably flexible way to represent probability distributions.

Actually, I lied: the product of factors is not yet a probability because the probabilities of all pictures must add up to one, and there's no guarantee that the products of factors for all pictures will do so. We need to normalize them, meaning divide each product by the sum of all of them. The sum of all the normalized products is then guaranteed to be one because it's just a number divided by itself. The probability of a picture is thus the weighted sum of its features, exponentiated and normalized. If you look back at the equation in the five-pointed star, you'll probably start to get an inkling of what it means. P is a probability, w is a vector of weights (notice it's in boldface), n is a vector of numbers, and their dot product • is exponentiated and divided by Z, the sum of all products. If we let the first component of n be one if the first feature of the image is true and zero otherwise, and so on, $w•n$ is just a shorthand for the weighted sum of features we've been talking about all along.

So the equation gives the probability of an image (or whatever) according to a Markov network. But it's more general than that because it's not just the equation of a Markov network; rather, it's the equation of a Markov logic network, as we call it. In a Markov logic network, or MLN for short, the numbers in n don't have to be just zero or one, and they don't refer to features—they refer to logical formulas. At the end of Chapter 8, we saw how we can go beyond Markov networks to relational models, which are defined in terms of feature templates, not just features. *Alice and Bob both have the flu* is a feature specific to Alice and Bob. *X and Y both have the flu* is a feature template, which can be instantiated with Alice and Bob, Alice and Chris, and any other two people. A feature template is a powerful thing because it can summarize billions of features or more in a single short expression. But we need a formal language to define feature templates, and we have one readily available: logic.

An MLN is just a set of logical formulas and their weights. When applied to a particular set of entities, it defines a Markov network over their possible states. For example, if the entities are Alice and Bob, a possible state is that Alice and Bob are friends, Alice has the flu, and so does Bob. Let's suppose the MLN has two formulas: *Everyone has the flu* and *If someone has the flu, so do their friends.* In standard logic, this would be a pretty useless pair of statements: the first would rule out any state with even a single healthy person, and the second would be redundant. But in an MLN, the first formula just means that there's a feature *X has the flu* for every person X, with the same weight as the formula. If people are likely to have the flu, the formula will have a high weight, and so will the corresponding features. A state with many healthy people is less probable than one with few, but not impossible. And because of the second formula, a state where someone has the flu and their friends don't is less probable than one where healthy and infected people fall into separate clusters of friends.

At this point you can probably guess what the n in the master equation is: its first component is the number of true instances of the first formula in the state, the second is the number of true instances of the second formula, and so on. If we're looking at a group of ten friends and seven of them have the flu, the first component of n is seven, and so on. (Shouldn't the probability be different if seven out of twenty instead of seven out of ten friends have the flu? Yes, and it is, because of Z.) In the limit, if we let all the weights go to infinity, Markov logic reduces to standard logic because violating a single instance of a formula then causes the probability to collapse to zero, making the state impossible. On the probabilistic side, an MLN reduces to a Markov network when all the formulas talk about a single object. So Markov logic includes both logic and Markov networks as special cases, and it's the unification we were looking for.

Learning an MLN means discovering formulas that are true in the world more often than random chance would predict, and figuring out the weights for those formulas that cause their predicted probabilities to match their observed frequencies. Once we've learned an MLN, we can use it to answer questions like "What is the probability that Bob has the flu, given that he's friends with Alice and she has the flu?" And guess what? It turns out that the probability is given by an S curve applied to the weighted sum of features, much as in a multilayer perceptron. And an MLN with long chains of rules can represent a deep neural network, with one layer per link in the chain.

Of course, don't be deceived by the simple MLN above for predicting the spread of flu. Picture instead an MLN for diagnosing and curing cancer. The MLN represents a probability distribution over the states of a cell. Every part of the cell, every organelle, every metabolic pathway, every gene and protein is an entity in the MLN, and the MLN's formulas encode the dependencies between them. We can ask the MLN, "Is this cell cancerous?" and probe it with different drugs and see what happens. We don't have an MLN like this yet, but later in this chapter I'll envisage how it might come about.

To recap: the unified learner we've arrived at uses MLNs as the representation, posterior probability as the evaluation function, and genetic search coupled with gradient descent as the optimizer. If we want, we can easily replace the posterior by some other accuracy measure, or genetic search by hill climbing. We've ascended a high peak, and now we can enjoy the view. I wouldn't be so rash as to call this learner the Master Algorithm, however. For one, the proof of the pudding is in the eating, and although over the last decade this algorithm (or variations of it) has been successfully applied in many areas, there are many more to which it hasn't, and so it's not yet clear just how general purpose it is. Second, there are some important problems that it doesn't solve. But before we look at them, let's look at what it can do.

From Hume to your housebot

You can download the learner I've just described from alchemy.cs .washington.edu. We christened it Alchemy to remind ourselves that, despite all its successes, machine learning is still in the alchemy stage of science. If you do download it, you'll see that it includes a lot more than the basic algorithm I've described but also that it is still missing a few things I said the universal learner ought to have, like crossover. Nevertheless, let's use the name Alchemy to refer to our candidate universal learner for simplicity.

Alchemy addresses Hume's original question by having another input besides the data: your initial knowledge, in the form of a set of logical formulas, with or without weights. The formulas can be inconsistent, incomplete, or even just plain wrong; the learning and probabilistic reasoning will take care of that. The key point is that Alchemy doesn't have to learn from scratch. In fact, we can even tell Alchemy to keep the formulas unchanged and learn only the weights. In this case, giving Alchemy the appropriate formulas can turn it into a Boltzmann machine, a Bayesian network, an instance-based learner, and many other models. This explains why we can have a universal learner despite the "no free lunch" theorem. Rather, Alchemy is like an inductive Turing machine, which we can program to behave as a very powerful or a very restricted learner; it's up to us. Alchemy provides a unifier for machine learning in the same way that the Internet provides one for computer networks, the relational model for databases, or the graphical user interface for everyday applications.

Of course, even if you use Alchemy with no initial formulas (and you can), that doesn't make it knowledge-free. The choice of formal language, score function, and optimizer implicitly encodes assumptions about the world. So it's natural to ask whether we can have an even more general learner than Alchemy. What did evolution assume when it began its long journey from the first bacteria to all the life-forms around today? I think there's a simple assumption from which all else follows: the learner is part of the world. This means that the learner as a physical system obeys the same laws as its environment, whatever they are, and therefore already "knows" them implicitly and is primed to discover them. In the next section, we'll see what this can mean concretely and how to embody it in Alchemy. But for the moment, let's note that it's perhaps the best answer we can ever give to Hume's question. On the one hand, assuming the learner is part of the world *is* an assumption—in principle, the learner could obey different laws from those the world obeys—so it satisfies Hume's dictum that learning is only possible with prior knowledge. On the other hand, it's an assumption so basic and hard to disagree with that perhaps it's all we need for this world.

At the other extreme, knowledge engineers—the most determined critics of machine learning—have good reason to like Alchemy. Instead of a basic model structure or a few rough guesses, Alchemy can input a large, lovingly assembled knowledge base, if it's available. Because probabilistic rules can interact in much richer ways than deterministic ones, manually encoded knowledge goes a longer way in Markov logic. And since knowledge bases in Markov logic don't have to be self-consistent, they can be very large and accommodate many different contributors without falling apart—a goal that has so far eluded knowledge engineers.

Most of all, though, Alchemy addresses the problems that each of the five tribes of machine learning has worked on for so long. Let's look at each of them in turn.

Symbolists combine different pieces of knowledge on the fly, in the same way that mathematicians combine axioms to prove theorems. This contrasts sharply with neural networks and other models with a fixed structure. Alchemy does it using logic, as symbolists do, but with a twist. To prove a theorem in logic, you need to find only one sequence of axiom applications that produces it. Because Alchemy reasons probabilistically, it does more: it finds multiple sequences of formulas that lead to the theorem or its negation and weighs them to compute the theorem's probability of being true. This way it can reason not just about mathematical universals, but about whether "the president" in a news story means "Barack Obama," or what folder an e-mail should be filed in. The symbolists' master algorithm, inverse deduction, postulates new logical rules needed to serve as steps between the data and a desired conclusion. Alchemy introduces new rules by hill climbing, starting with the initial rules and constructing rules that, combined with the initial ones and the data, make the conclusions more likely.

Connectionists' models are inspired by the brain, with networks of S curves that correspond to neurons and weighted connections between them corresponding to synapses. In Alchemy, two variables are connected if they appear together in some formula, and the probability of a variable given its neighbors is an S curve. (Although I won't show why, it's a direct consequence of the master equation we saw in the previous section.) The connectionists' master algorithm is backpropagation, which they use to figure out which neurons are responsible for which errors and adjust their weights accordingly. Backpropagation is a form of gradient descent, which Alchemy uses to optimize the weights of a Markov logic network.

Evolutionaries use genetic algorithms to simulate natural selection. A genetic algorithm maintains a population of hypotheses and in each generation crosses over and mutates the fittest ones to produce the next generation. Alchemy maintains a population of hypotheses in the form of weighted formulas, modifies them in various ways at each step, and keeps the variations that most increase the posterior probability of the data (or some other score function). If the population is a single hypothesis, this reduces to hill climbing. The current open-source implementation of Alchemy does not include crossover, but this would be a straightforward addition. The evolutionaries' master algorithm is genetic programming, which applies crossover and mutation to computer programs represented as trees of subroutines. Trees of subroutines can be represented by sets of logical rules, and the Prolog programming language does just that. In Prolog, each rule corresponds to a subroutine, and its antecedents are the subroutines it calls. So we can think of Alchemy with crossover as genetic programming using a Prolog-like programming language, with the added advantage that the rules can be probabilistic.

Bayesians believe that modeling uncertainty is the key to learning and use formal representations like Bayesian networks and Markov networks to do so. As we already saw, Markov networks are a special type of MLN. Bayesian networks are also easily represented using the MLN master equation, with a feature for each possible state of a variable and its parents, and the logarithm of the corresponding conditional probability as its weight. (The normalization constant Z then conveniently reduces to 1, meaning we can ignore it.) Bayesians' master algorithm is Bayes' theorem, implemented using probabilistic inference algorithms like belief propagation and MCMC. As you may have noticed, Bayes' theorem is a special case of the master equation, with $P = P(A|B)$, $Z = P(B)$, and features and weights corresponding to $P(A)$ and $P(B|A)$. The Alchemy system includes both belief propagation and MCMC for inference, generalized to handle weighted logical formulas. Using probabilistic inference over the proof paths provided by logic, Alchemy weighs the evidence for and against a conclusion and outputs the probability of the conclusion. This contrasts with the "plain vanilla" logic used by symbolists, which is all or none and so falls apart when given contradictory evidence.

Analogizers learn by hypothesizing that entities with similar known properties have similar unknown ones: patients with similar symptoms have similar diagnoses, readers who bought the same books in the past will do so again in the future, and so on. MLNs can represent similarity between entities with formulas like *People with the same tastes buy the same books*. Then the more of the same books Alice and Bob have bought, the more likely they are to have the same tastes, and (applying the same formula in the opposite direction) the more likely Alice is to buy a book if Bob also did. Their similarity is represented by their probability of having the same tastes. To make this really useful, we can have different weights for different instances of the same rule: if Alice and Bob both bought a certain rare book, this is probably more informative than if they both bought a best seller and should therefore have a higher weight. In this case the properties whose similarity we're computing are discrete (bought/not bought), but we can also represent similarity between continuous properties, like the distance between two cities, by letting an MLN have these similarities as features. If the evaluation function is a margin-style score function instead of the posterior probability, the result is a generalization of SVMs, the analogizers' master algorithm. A greater challenge for our master learner is reproducing structure mapping, the more powerful type of analogy that can make inferences from one domain (e.g., the solar system) to another (the atom). We can do this by learning formulas that don't refer to any of the specific relations in the source domain. For example, *Friends of smokers also smoke* is about friendship and smoking, but *Related entities have similar properties* applies to any relation and property. We can learn it by generalizing from *Friends of friends also smoke, Coworkers of experts are also experts,* and other such patterns in a social network and then apply it to, say, the web, with instances like *Interesting pages link to interesting pages,* or to molecular biology, with instances like *Proteins that interact with gene-regulating proteins also regulate genes.* Researchers in my group and others have done all of these things, and more.

Alchemy also enables the five types of unsupervised learning we saw in the previous chapter. It does relational learning, obviously, and in fact that's where most of its applications to date have been. Alchemy uses logic to represent relations among entities and Markov networks to let them be uncertain. We can turn Alchemy into a reinforcement learner by wrapping delayed rewards around it and using it to learn the value of each state in the same way that traditional reinforcement learners use, say, a neural network. We can do chunking in Alchemy by adding a new operation that condenses chains of rules into single rules. (For example, *If A then B* and *If B then C* into *If A then C.*) An MLN with a single unobserved variable connected to all the observable ones does clustering. (An unobserved variable is a variable whose values we never see in the data; it's "hidden," so to speak, and can only be inferred.) MLNs with more than one unobserved variable do a kind of discrete dimensionality reduction by inferring the values of those (fewer) variables from the (more numerous) observable ones. Alchemy can also handle MLNs with continuous unobserved variables, which would be needed to do things like principal-component analysis and Isomap. So Alchemy can in principle do all the things we want Robby the robot to do, or at least all the things we've discussed in this book. Indeed, we've used Alchemy to let a robot learn a map of its environment, figuring out from its sensors where the walls and doors are, their angles and distances, and so on, which is the first step in building a competent housebot.

Finally, we can turn Alchemy into a metalearner like stacking by encoding the individual classifiers as MLNs and adding or learning formulas to combine them. This is what DARPA did in its PAL project. PAL, the Personalized Assistant that Learns, was the largest AI project in DARPA history and the progenitor of Siri. PAL's goal was to build an automated secretary. It used Markov logic as its overarching representation, combining the outputs from different modules into the final decisions on what to do. This also allowed PAL's modules to learn from each other by evolving toward a consensus.

One of Alchemy's largest applications to date was to learn a semantic network (or knowledge graph, as Google calls it) from the web. A semantic network is a set of concepts (like planets and stars) and relations among those concepts (planets orbit stars). Alchemy learned over a million such patterns from facts extracted from the web (e.g., Earth orbits the sun). It discovered concepts like planet all by itself. The version we used was more advanced than the basic one I've described here, but the essential ideas are the same. Various research groups have used Alchemy or their own MLN implementations to solve problems in natural language processing, computer vision, activity recognition, social network analysis, molecular biology, and many other areas.

Despite its successes, Alchemy has some significant shortcomings. It does not yet scale to truly big data, and someone without a PhD in machine learning will find it hard to use. Because of these problems, it's not yet ready for prime time. But let's see what we can do about them.

Planetary-scale machine learning

In computer science, a problem isn't really solved until it's solved efficiently. Knowing how to do something isn't much use if you can't do it within the available time and memory, and these can run out very quickly when you're dealing with an MLN. We routinely learn MLNs with millions of variables and billions of features, but this is not as large as it seems because the number of variables grows very quickly with the number of entities in the MLN: if you have a social network with a thousand people, you already have a million possible pairs of friends and a billion instances of the formula *Friends of friends are friends.*

Inference in Alchemy is a combination of logical and probabilistic inference. The former is done by proving theorems and the latter by belief propagation, MCMC, and the other methods we saw in Chapter 6. We've combined the two into probabilistic theorem proving, and the unified inference algorithm, capable of computing the probability of any logical formula, is a key part of the current Alchemy system. But it can be very computationally expensive. If your brain used probabilistic theorem proving, the proverbial tiger would eat you before you figured out to run away. That's a high price to pay for the generality of Markov logic. Your brain, having evolved in the real world, must encode additional assumptions that allow it to do inference very efficiently. In the last few years, we've started to figure out what they might be and encode them into Alchemy.

The world is not a random jumble of interactions; it has a hierarchical structure: galaxies, planets, continents, countries, cities, neighborhoods, your house, you, your head, your nose, a cell on its tip, the organelles in it, molecules, atoms, subatomic particles. The way to model it, then, is with an MLN that also has a hierarchical structure. This is an example of the assumption that the learner and its environment are alike. The MLN doesn't have to know a priori which parts the world is composed of; all Alchemy has to do is assume that the world *has* parts and look for them, rather like a newly made bookshelf assumes that there are books but doesn't yet know which ones will be placed on it. Hierarchical structure helps make inference tractable because subparts of the world interact mostly with other subparts of the same part: neighbors talk more to each other than to people in another country, molecules produced in one cell react mostly with other molecules in that cell, and so on.

Another property of the world that makes learning and inference easier is that the entities in it don't come in arbitrary forms. Rather, they fall into classes and subclasses, with members of the same class being more alike than members of different ones. Alive or inanimate, animal or plant, bird or mammal, human or not: if we know all the distinctions relevant to the question at hand, we can lump together all the entities that lack them and that can save a lot of time. As before, the MLN doesn't have to know a priori what the classes in the world are; it can learn them from data by hierarchical clustering.

The world has parts, and parts belong to classes: combining these two gives us most of what we need to make inference in Alchemy tractable. We can learn the world's MLN by breaking it into parts and subparts, such that most interactions are between subparts of the same part, and then grouping the parts into classes and subclasses. If the world is a Lego toy, we can break it up into individual bricks, remembering which attaches to which, and group the bricks by shape and color. If the world is Wikipedia, we can extract the entities it talks about, group them into classes, and learn how classes relate to each other. Then if someone asks us "Is Arnold Schwarzenegger an action star?" we can answer yes, because he's a star and he's in action movies. Step-by-step, we can learn larger and larger MLNs, until we're doing what a friend of mine at Google calls "planetary-scale machine learning": modeling everyone in the world at once, with data continually streaming in and answers streaming out.

Of course, learning on this scale requires much more than a direct implementation of the algorithms we've seen. For one, beyond a certain point a single processor is not enough; we have to distribute the learning over many servers. Researchers in both industry and academia have intensely investigated how to, for example, do gradient descent using many computers in parallel. One option is to divide the data among the processors; another is to divide the model's parameters. After each step, we combine the results and redistribute the work. Either way, doing this without letting the cost of communication overwhelm you, or the quality of the results suffer, is far from trivial. Another issue is that, if you have an endless stream of data coming in, you can't wait to see it all before you commit to some decisions. One solution is to use the sampling principle: if you want to predict who will win the next presidential election, you don't need to ask every voter who he or she will vote for; a sample of a few thousand suffices, if you're willing to accept a little bit of uncertainty. The trick is to generalize this to complex models with millions of parameters. But we can do this by taking at each step just as many examples from the stream as we need to be pretty sure that we're making the right decision and that the total uncertainty over all the decisions stays within bounds. That way we can effectively learn from infinite data in finite time, as I put it in an early paper proposing this approach.

Big-data systems are the Cecil B. DeMille productions of machine learning, with thousands of servers instead of thousands of extras. In the largest projects, just getting all the data together, verifying it, cleaning it up, and munging it into a form the learners can digest can make building the pyramids seem like a walk in the park. At the pharaonic end, Europe's FuturICT project aims to build a model of—literally—the whole world. Societies, governments, culture, technology, agriculture, disease, the global economy: nothing is to be left out. This is surely premature, but it does foreshadow the shape of things to come. In the meantime, projects like this can help us find out where the limits of scalability are and how to overcome them.

Computational complexity is one thing, but human complexity is another. If computers are like idiot savants, learning algorithms can sometimes come across like child prodigies prone to temper tantrums. That's one reason humans who can wrangle them into submission are so highly paid. If you know how to expertly tweak the control knobs until they're just right, magic can ensue, in the form of a stream of insights beyond the learner's years. And, not unlike the Delphic oracle, interpreting the learner's pronouncements can itself require considerable skill. Turn the knobs wrong, though, and the learner may spew out a torrent of gibberish or clam up in defiance. Unfortunately, in this regard Alchemy is no better than most. Writing down what you know in logic, feeding in the data, and pushing the button is the fun part. When Alchemy returns a beautifully accurate and efficient MLN, you go down to the pub and celebrate. When it doesn't—which is most of the time—the battle begins. Is the problem in the knowledge, the learning, or the inference? On the one hand, because of the learning and probabilistic inference, a simple MLN can do the job of a complex program. On the other, when it doesn't work, it's much harder to debug. The solution is to make it more interactive, able to introspect and explain its reasoning. That will take us another step closer to the Master Algorithm.

The doctor will see you now

The cure for cancer is a program that inputs the cancer's genome and outputs the drug to kill it with. We can now picture what such a program—let's call it CanceRx—will look like. Despite its outward simplicity, CanceRx is one of the largest and most complex programs ever built—indeed, so large and complex that it could only have been built with the help of machine learning. It is based on a detailed model of how living cells work, with a subclass for each type of cell in the human body and an overarching model of how they interact. This model, in the form of an MLN or something akin to it, combines knowledge of molecular biology with vast amounts of data from DNA sequencers, microarrays, and many other sources. Some of the knowledge was manually encoded, but most was automatically extracted from the biomedical literature. The model is continually evolving, incorporating the results of new experiments, data sources, and patient histories. Ultimately, it will know every pathway, regulatory mechanism, and chemical reaction in every type of human cell—the sum total of human molecular biology.

CanceRx spends most of its time querying the model with candidate drugs. Given a new drug, the model predicts its effect on both cancer cells and normal ones. When Alice is diagnosed with cancer, CanceRx instantiates its model with both her normal cells and the tumor's and tries all available drugs until it finds one that kills the cancer cells without harming the healthy ones. If it can't find a drug or combination of drugs that works, it sets about designing one that will, perhaps evolving it from existing ones using hill climbing or crossover. At each step in the search, it tries the candidate drugs on the model. If a drug stops the cancer but still has some harmful side effect, CanceRx tries to tweak it to get rid of the side effect. When Alice's cancer mutates, it repeats the whole process. Even before the cancer mutates, the model predicts likely mutations, and CanceRx prescribes drugs that will stop them dead in their tracks. In the game of chess between humanity and cancer, CanceRx is checkmate.

Notice that machine learning isn't going to give us CanceRx all by itself. It's not as if we have a vast database of molecular biology ready to go, stream it into the Master Algorithm, and out pops the perfect model of a living cell. CanceRx would be the end result, after many iterations, of a worldwide collaboration between hundreds of thousands of biologists, oncologists, and data scientists. Most important, however, CanceRx would incorporate data from millions of cancer patients, with the help of their doctors and hospitals. Without that data, we can't cure cancer; with it, we can. Contributing to this growing database would not only be in every cancer patient's interest; it would be her ethical duty. In the world of CanceRx, discrete clinical trials are a thing of the past; new treatments proposed by CanceRx are continually being rolled out, and if they work, given to a widening circle of patients. Both successes and failures provide valuable data for CanceRx's learning, in a virtuous circle of improvement. If you look at it one way, machine learning is only a small part of the CanceRx project, well behind data gathering and human contributions. But looked at another way, machine learning is the linchpin of the whole enterprise. Without it, we would have only fragmentary knowledge of cancer biology, scattered among thousands of databases and millions of scientific articles, each doctor aware of only a small part. Assembling all this knowledge into a coherent whole is beyond the power of unaided humans, no matter how smart; only machine learning can do it. Because every cancer is different, it takes machine learning to find the common patterns. And because a single tissue can yield billions of data points, it takes machine learning to figure out what to do for each new patient.

The effort to build what will ultimately become CanceRx is already under way. Researchers in the new field of systems biology model whole metabolic networks rather than individual genes and proteins. One group at Stanford has built a model of a whole cell. The Global Alliance for Genomics and Health promotes data sharing among researchers and oncologists, with a view to large-scale analysis. CancerCommons .org assembles cancer models and lets patients pool their histories and learn from similar cases. Foundation Medicine pinpoints the mutations in a patient's tumor cells and suggests the most appropriate drugs. A decade ago, it wasn't clear if, or how, cancer would ever be cured. Now we can see how to get there. The road is long, but we have found it.

CHAPTER TEN

This Is the World on Machine Learning

Now that you've toured the machine learning wonderland, let's switch gears and see what it all means to you. Like the red pill in *The Matrix*, the Master Algorithm is the gateway to a different reality: the one you already live in but didn't know it yet. From dating to work, from self-knowledge to the future of society, from data sharing to war, and from the dangers of AI to the next step in evolution, a new world is taking shape, and machine learning is the key that unlocks it. This chapter will help you make the most of it in your life and be ready for what comes next. Machine learning will not single-handedly determine the future, any more than any other technology; it's what we decide to do with it that counts, and now you have the tools to decide.

Chief among these tools is the Master Algorithm. Whether it arrives sooner or later, and whether or not it looks like Alchemy, is less important than what it encapsulates: the essential capabilities of a learning algorithm, and where they'll take us. We can equally well think of the Master Algorithm as a composite picture of current and future learners, which we can conveniently use in our thought experiments in lieu of the specific algorithm inside product X or website Y, which the respective companies are unlikely to share with us anyway. Seen in this light, the learners we interact with every day are embryonic versions of the Master Algorithm, and our task is to understand them and shape their growth to better serve our needs.

In the coming decades, machine learning will affect such a broad swath of human life that one chapter of one book cannot possibly do it justice. Nevertheless, we can already see a number of recurring themes, and it's those we'll focus on, starting with what psychologists call theory of mind—the computer's theory of your mind, that is.

Sex, lies, and machine learning

Your digital future begins with a realization: every time you interact with a computer—whether it's your smart phone or a server thousands of miles away—you do so on two levels. The first one is getting what you want there and then: an answer to a question, a product you want to buy, a new credit card. The second level, and in the long run the most important one, is teaching the computer about you. The more you teach it, the better it can serve you—or manipulate you. Life is a game between you and the learners that surround you. You can refuse to play, but then you'll have to live a twentieth-century life in the twenty-first. Or you can play to win. What model of you do you want the computer to have? And what data can you give it that will produce that model? Those two questions should always be in the back of your mind whenever you interact with a learning algorithm—as they are when you interact with other people. Alice knows that Bob has a mental model of her and seeks to shape it through her behavior. If Bob is her boss, she tries to come across as competent, loyal, and hardworking. If instead Bob is someone she's trying to seduce, she'll be at her most seductive. We could hardly function in society without this ability to intuit and respond to what's on other people's minds. The novelty in the world today is that computers, not just people, are starting to have theories of mind. Their theories are still primitive, but they're evolving quickly, and they're what we have to work with to get what we want—no less than with other people. And so *you* need a theory of *the computer's* mind, and that's what the Master Algorithm provides, after plugging in the score function (what you think the learner's goals are, or more precisely its owner's) and the data (what you think it knows).

Take online dating. When you use Match.com, eHarmony, or Ok-Cupid (suspend your disbelief, if necessary), your goal is simple: to find the best possible date you can. But chances are it will take a lot of work and several disappointing dates before you meet someone you really like. One hardy geek extracted twenty thousand profiles from OkCupid, did his own data mining, found the woman of his dreams on the eighty-eighth date, and told his odyssey to *Wired* magazine. To succeed with fewer dates and less work, your two main tools are your profile and your responses to suggested matches. One popular option is to lie (about your age, for example). This may seem unethical, not to mention liable to blow up in your face when your date discovers the truth, but there's a twist. Savvy online daters already know that people lie about their age on their profiles and adjust accordingly, so if you state your true age, you're effectively telling them you're older than you really are! In turn, the learner doing the matching thinks people prefer younger dates than they really do. The logical next step is for people to lie about their age by even more, ultimately rendering this attribute meaningless.

A better way for all concerned is to focus on your specific, unusual attributes that are highly predictive of a match, in the sense that they pick out people you like that not everyone else does, and therefore have less competition for. Your job (and your prospective date's) is to provide these attributes. The matcher's job is to learn from them, in the same way that an old-fashioned matchmaker would. Compared to a village matchmaker, Match.com's algorithm has the advantage that it knows vastly more people, but the disadvantage is that it knows them much more superficially. A naïve learner, such as a perceptron, will be content with broad generalizations like "gentlemen prefer blondes." A more sophisticated one will find patterns like "people with the same unusual musical tastes are often good matches." If Alice and Bob both like Beyoncé, that alone hardly singles them out for each other. But if they both like Bishop Allen, that makes them at least a little bit more likely to be potential soul mates. If they're both fans of a band the learner does not know about, that's even better, but only a relational algorithm like Alchemy can pick it up. The better the learner, the more it's worth your time to teach it about you. But as a rule of thumb, you want to differentiate yourself enough so that it won't confuse you with the "average person" (remember Bob Burns from Chapter 8), but not be so unusual that it can't fathom you.

Online dating is in fact a tough example because chemistry is hard to predict. Two people who hit it off on a date may wind up falling in love and believing passionately that they were made for each other, but if their initial conversation takes a different turn, they might instead find each other annoying and never want to meet again. What a really sophisticated learner would do is run a thousand Monte Carlo simulations of a date between each pair of plausible matches and rank the matches by the fraction of dates that turned out well. Short of that, dating sites can organize parties and invite people who are each a likely match for many of the others, letting them accomplish in a few hours what would otherwise take weeks.

For those of us who are not keen on online dating, a more immediately useful notion is to choose which interactions to record and where. If you don't want your Christmas shopping to leave Amazon confused about your tastes, do it on other sites. (Sorry, Amazon.) If you watch different kinds of videos at home and for work, keep two accounts on YouTube, one for each, and YouTube will learn to make the corresponding recommendations. And if you're about to watch some videos of a kind that you ordinarily have no interest in, log out first. Use Chrome's incognito mode not for guilty browsing (which you'd never do, of course) but for when you don't want the current session to influence future personalization. On Netflix, adding profiles for the different people using your account will spare you R-rated recommendations on family movie night. If you don't like a company, click on their ads: this will not only waste their money now, but teach Google to waste it again in the future by showing the ads to people who are unlikely to buy the products. And if you have very specific queries that you want Google to answer correctly in the future, take a moment to trawl through the later results pages for the relevant links and click on them. More generally, if a system keeps recommending the wrong things to you, try teaching it by finding and clicking on a bunch of the right ones and come back later to see if it did.

That could be a lot of work, though. What all of these illustrate, unfortunately, is how narrow the communication channel between you and the learner is today. You should be able to tell it as much as you want about yourself, not just have it learn indirectly from what you do. More than that, you should be able to inspect the learner's model of you and correct it as desired. The learner can still decide to ignore you, if it thinks you're lying or are low on self-knowledge, but at least it would be able to take your input into account. For this, the model needs to be in a form that humans understand, such as a set of rules rather than a neural network, and it needs to accept general statements as input in addition to raw data, as Alchemy does. All of which brings us to the question of how good a model of you a learner can have and what you'd want to do with that model.

The digital mirror

Take a moment to consider all the data about you that's recorded on all the world's computers: your e-mails, Office docs, texts, tweets, and Facebook and LinkedIn accounts; your web searches, clicks, downloads, and purchases; your credit, tax, phone, and health records; your Fitbit statistics; your driving as recorded by your car's microprocessors; your wanderings as recorded by your cell phone; all the pictures of you ever taken; brief cameos on security cameras; your Google Glass snippets—and so on and so forth. If a future biographer had access to nothing but this "data exhaust" of yours, what picture of you would he form? Probably a quite accurate and detailed one in many ways, but also one where some essential things would be missing. Why did you, one beautiful day, decide to change careers? Could the biographer have predicted it ahead of time? What about that person you met one day and secretly never forgot? Could the biographer wind back through the found footage and say "Ah, there"?

The sobering (or perhaps reassuring) thought is that no learner in the world today has access to all this data (not even the NSA), and even if it did, it wouldn't know how to turn it into a real likeness of you. But suppose you took all your data and gave it to the—real, future—Master Algorithm, already seeded with everything we could teach it about human life. It would learn a model of you, and you could carry that model in a thumb drive in your pocket, inspect it at will, and use it for everything you pleased. It would surely be a wonderful tool for introspection, like looking at yourself in the mirror, but it would be a digital mirror that showed not just your looks but all things observable about you—a mirror that could come alive and converse with you. What would you ask it? Some of the answers you might not like, but that would be all the more reason to ponder them. And some would give you new ideas, new directions. The Master Algorithm's model of you might even help you become a better person.

Self-improvement aside, probably the first thing you'd want your model to do is negotiate the world on your behalf: let it loose in cyberspace, looking for all sorts of things for you. From all the world's books, it would suggest a dozen you might want to read next, with more insight than Amazon could dream of. Likewise for movies, music, games, clothes, electronics—you name it. It would keep your refrigerator stocked at all times, natch. It would filter your e-mail, voice mail, Facebook posts, and Twitter feed and, when appropriate, reply on your behalf. It would take care of all the little annoyances of modern life for you, like checking credit-card bills, disputing improper charges, making arrangements, renewing subscriptions, and filling out tax returns. It would find a remedy for your ailment, run it by your doctor, and order it from Walgreens. It would bring interesting job opportunities to your attention, propose vacation spots, suggest which candidates to vote for on the ballot, and screen potential dates. And, after the match was made, it would team up with your date's model to pick some restaurants you might both like. Which is where things start to get *really* interesting.

A society of models

In this rapidly approaching future, you're not going to be the only one with a "digital half" doing your bidding twenty-four hours a day. Everyone will have a detailed model of him- or herself, and these models will talk to each other all the time. If you're looking for a job and company X is looking to hire, its model will interview your model. It will be a lot like a real, flesh-and-blood interview—your model will still be well advised to not volunteer negative information about you, and so on—but it will take only a fraction of a second. You'll click on "Find Job" in your future LinkedIn account, and you'll immediately interview for every job in the universe that remotely fits your parameters (profession, location, pay, etc.). LinkedIn will respond on the spot with a ranked list of the best prospects, and out of those, you'll pick the first company that you want to have a chat with. Same with dating: your model will go on millions of dates so you don't have to, and come Saturday, you'll meet your top prospects at an OkCupid-organized party, knowing that you're also one of *their* top prospects—and knowing, of course, that their *other* top prospects are also in the room. It's sure to be an interesting night.

In the world of the Master Algorithm, "my people will call your people" becomes "my program will call your program." Everyone has an entourage of bots, smoothing his or her way through the world. Deals get pitched, terms negotiated, arrangements made, all before you lift a finger. Today, drug companies target your doctor, because he decides what drugs to prescribe to you. Tomorrow, the purveyors of every product and service you use, or might use, will target your model, because your model will screen them for you. Their bots' job is to get your bot to buy. Your bot's job is to see through their claims, just as you see through TV commercials, but at a much finer level of detail, one that you'd never have the time or patience for. Before you buy a car, the digital you will go over every one of its specs, discuss them with the manufacturer, and study everything anyone in the world has said about that car and its alternatives. Your digital half will be like power steering for your life: it goes where you want to go but with less effort from you. This does not mean that you'll end up in a "filter bubble," seeing only what you reliably like, with no room for the unexpected; the digital you knows better than that. Part of its brief is to leave some things open to chance, to expose you to new experiences, and to look for serendipity.

Even more interesting, the process doesn't end when you find a car, a house, a doctor, a date, or a job. Your digital half is continually learning from its experiences, just as you would. It figures out what works and doesn't, whether it's in job interviews, dating, or real-estate hunting. It learns about the people and organizations it interacts with on your behalf and then (even more important) from your real-world interactions with them. It predicted Alice would be a great date for you, but you had an awkward time, so it hypothesizes possible reasons, which it will test on your next round of dating. It shares its most important findings with you. ("You believe you like X, but in reality you tend to go for Y.") Comparing your experiences of various hotels with their reviews on TripAdvisor, it figures out what the really telling tidbits are and looks for them in the future. It learns not just which online merchants are more trustworthy but how to decode what the less trustworthy ones say. Your digital half has a model of the world: not just of the world in general but of the world as it relates to you. At the same time, of course, everyone else also has a continually evolving model of his or her world. Every party to an interaction learns from it and applies what it's learned to its next interactions. You have your model of every person and organization you interact with, and they each have their model of you. As the models improve, their interactions become more and more like the ones you would have in the real world—except millions of times faster and in silicon. Tomorrow's cyberspace will be a vast parallel world that selects only the most promising things to try out in the real one. It will be like a new, global subconscious, the collective id of the human race.

To share or not to share, and how and where

Of course, learning about the world all by yourself is slow, even if your digital half does it orders of magnitude faster than the flesh-and-blood you. If others learn about you faster than you learn about them, you're in trouble. The answer is to share: a million people learn about a company or product a lot faster than a single one does, provided they pool their experiences. But who should you share data with? That's perhaps the most important question of the twenty-first century.

Today your data can be of four kinds: data you share with everyone, data you share with friends or coworkers, data you share with various companies (wittingly or not), and data you don't share. The first type includes things like Yelp, Amazon, and TripAdvisor reviews, eBay feedback scores, LinkedIn résumés, blogs, tweets, and so on. This data is very valuable and is the least problematic of the four. You make it available to everyone because you want to, and everyone benefits. The only problem is that the companies hosting the data don't necessarily allow it to be downloaded in bulk for building models. They should. Today you can go to TripAdvisor and see the reviews and star ratings of particular hotels you're considering, but what about a model of what makes a hotel good or bad in general, which you could use to rate hotels that currently have few or no reliable reviews? TripAdvisor could learn it, but what about a model of what makes a hotel good or bad *for you*? This requires information about you that you may not want to share with TripAdvisor. What you'd like is a trusted party that combines the two types of data and gives you the results.

The second kind of data should also be unproblematic, but it isn't because it overlaps with the third. You share updates and pictures with your friends on Facebook, and they with you. But everyone shares their updates and pictures with Facebook. Lucky Facebook: it has a billion friends. Day by day, it learns a lot more about the world than any one person does. It would learn even more if it had better algorithms, and they are getting better every day, courtesy of us data scientists. Facebook's main use for all this knowledge is to target ads to you. In return, it provides the infrastructure for your sharing. That's the bargain you make when you use Facebook. As its learning algorithms improve, it gets more and more value out of the data, and some of that value returns to you in the form of more relevant ads and better service. The only problem is that Facebook is also free to do things with the data and the models that are not in your interest, and you have no way to stop it.

This problem pops up across the board with data you share with companies, which these days includes pretty much everything you do online as well as a lot of what you do offline. In case you haven't noticed, there's a mad race to gather data about you. Everybody loves your data, and no wonder: it's the gateway to your world, your money, your vote, even your heart. But everyone has only a sliver of it. Google sees your searches, Amazon your online purchases, AT&T your phone calls, Apple your music downloads, Safeway your groceries, Capital One your credit-card transactions. Companies like Acxiom collate and sell information about you, but if you inspect it (which in Acxiom's case you can, at aboutthedata.com), it's not much, and some of it is wrong. No one has anything even approaching a complete picture of you. That's both good and bad. Good because if someone did, they'd have far too much power. Bad because as long as that's the case there can be no 360-degree model of you. What you really want is a digital you that you're the sole owner of and that others can access only on your terms.

The last type of data—data you don't share—also has a problem, which is that maybe you should share it. Maybe it hasn't occurred to you to do so, maybe there's no easy way to, or maybe you just don't want to. In the latter case, you should consider whether you have an ethical responsibility to share. One example we've seen is cancer patients, who can contribute to curing cancer by sharing their tumors' genomes and treatment histories. But it goes well beyond that. All sorts of questions about society and policy can potentially be answered by learning from the data we generate in our daily lives. Social science is entering a golden age, where it finally has data commensurate with the complexity of the phenomena it studies, and the benefits to all of us could be enormous—provided the data is accessible to researchers, policy makers, and citizens. This does not mean letting others peek into your private life; it means letting them see the learned models, which should contain only statistical information. So between you and them there needs to be an honest data broker that guarantees your data won't be misused, but also that no free riders share the benefits without sharing the data.

In sum, all four kinds of data sharing have problems. These problems all have a common solution: a new type of company that is to your data what your bank is to your money. Banks don't steal your money (with rare exceptions). They're supposed to invest it wisely, and your deposits are FDIC-insured. Many companies today offer to consolidate your data somewhere in the cloud, but they're still a far cry from your personal data bank. If they're cloud providers, they try to lock you in—a big no-no. (Imagine depositing your money with Bank of America and not knowing if you'll be able to transfer it to Wells Fargo somewhere down the line.) Some startups offer to hoard your data and then mete it out to advertisers in return for discounts, but to me that misses the point. Sometimes you want to give information to advertisers for free because it's in your interests, sometimes you don't want to give it at all, and what to share when is a problem that only a good model of you can solve.

The kind of company I'm envisaging would do several things in return for a subscription fee. It would anonymize your online interactions, routing them through its servers and aggregating them with its other users'. It would store all the data from all your life in one place—down to your 24/7 Google Glass video stream, if you ever get one. It would learn a complete model of you and your world and continually update it. And it would use the model on your behalf, always doing exactly what you would, to the best of the model's ability. The company's basic commitment to you is that your data and your model will never be used against your interests. Such a guarantee can never be foolproof—you yourself are not guaranteed to never do anything against your interests, after all. But the company's life would depend on it as much as a bank's depends on the guarantee that it won't lose your money, so you should be able to trust it as much as you trust your bank.

A company like this could quickly become one of the most valuable in the world. As Alexis Madrigal of the *Atlantic* points out, today your profile can be bought for half a cent or less, but the value of a user to the Internet advertising industry is more like $1,200 per year. Google's sliver of your data is worth about $20, Facebook's $5, and so on. Add to that all the slivers that no one has yet, and the fact that the whole is more than the sum of the parts—a model of you based on all your data is much better than a thousand models based on a thousand slivers—and we're looking at easily over a trillion dollars per year for an economy the size of the United States. It doesn't take a large cut of that to make a Fortune 500 company. If you decide to take up the challenge and wind up becoming a billionaire, remember where you first got the idea.

Of course, some existing companies would love to host the digital you. Google, for example. Sergey Brin says that "we want Google to be the third half of your brain," and some of Google's acquisitions are probably not unrelated to how well their streams of user data complement its own. But, despite their head start, companies like Google and Facebook are not well suited to being your digital home because they have a conflict of interest. They earn a living by targeting ads, and so they have to balance your interests and the advertisers'. You wouldn't let the first or second half of your brain have divided loyalties, so why would you let the third?

One possible showstopper is that the government may subpoena your data or even preventively jail you, *Minority Report*–style, if your model looks like a criminal's. To forestall that, your data company can keep everything encrypted, with the key in your possession. (These days you can even compute over encrypted data without ever decrypting it.) Or you can keep it all in your hard disk at home, and the company just rents you the software.

If you don't like the idea of a profit-making entity holding the keys to your kingdom, you can join a data union instead. (If there isn't one in your neck of the cyberwoods yet, consider starting it.) The twentieth century needed labor unions to balance the power of workers and bosses. The twenty-first needs data unions for a similar reason. Corporations have a vastly greater ability to gather and use data than individuals. This leads to an asymmetry in power, and the more valuable the data—the better and more useful the models that can be learned from it—the greater the asymmetry. A data union lets its members bargain on equal terms with companies about the use of their data. Perhaps labor unions can get the ball rolling, and shore up their membership, by starting data unions for their members. But labor unions are organized by occupation and location; data unions can be more flexible. Join up with people you have a lot in common with; the models learned will be more useful to you that way. Notice that being in a data union does not mean letting other members see your data; it just means letting everyone use the models learned from the pooled data. Data unions can also be your vehicle for telling politicians what you want. Your data can influence the world as much as your vote—or more—because you only go to the polls on election day. On all other days, your data is your vote. Stand up and be counted!

So far I haven't uttered the word *privacy*. That's not by accident. Privacy is only one aspect of the larger issue of data sharing, and if we focus on it to the detriment of the whole, as much of the debate to date has, we risk reaching the wrong conclusions. For example, laws that forbid using data for any purpose other than the originally intended one are extremely myopic. (Not a single chapter of *Freakonomics* could have been written under such a law.) When people have to trade off privacy against other benefits, as when filling out a profile on a website, the implied value of privacy that comes out is much lower than if you ask them abstract questions like "Do you care about your privacy?" But privacy debates are more often framed in terms of the latter. The European Union's Court of Justice has decreed that people have the right to be forgotten, but they also have the right to remember, whether it's with their neurons or a hard disk. So do companies, and up to a point, the interests of users, data gatherers, and advertisers are aligned. Wasted attention benefits no one, and better data makes better products. Privacy is not a zero-sum game, even though it's often treated like one.

Companies that host the digital you and data unions are what a mature future of data in society looks like to me. Whether we'll get there is an open question. Today, most people are unaware of both how much data about them is being gathered and what the potential costs and benefits are. Companies seem content to continue doing it under the radar, terrified of a blowup. But sooner or later a blowup will happen, and in the ensuing fracas, draconian laws will be passed that in the end will serve no one. Better to foster awareness now and let everyone make their individual choices about what to share, what not, and how and where.

A neural network stole my job

How much of your brain does your job use? The more it does, the safer you are. In the early days of AI, the common view was that computers would replace blue-collar workers before white-collar ones, because white-collar work requires more brains. But that's not quite how things turned out. Robots assemble cars, but they haven't replaced construction workers. On the other hand, machine-learning algorithms have replaced credit analysts and direct marketers. As it turns out, evaluating credit applications is easier for machines than walking around a construction site without tripping, even though for humans it's the other way around. The common theme is that narrowly defined tasks are easily learned from data, but tasks that require a broad combination of skills and knowledge aren't. Most of your brain is devoted to vision and motion, which is a sign that walking around is much more complex than it seems; we just take it for granted because, having been honed to perfection by evolution, it's mostly done subconsciously. The company Narrative Science has an AI system that can write pretty good summaries of baseball games, but not novels, because—*pace* George F. Will—there's a lot more to life than to baseball games. Speech recognition is hard for computers because it's hard to fill in the blanks—literally, the sounds speakers routinely elide—when you have no idea what the person is talking about. Algorithms can predict stock fluctuations but have no clue how they relate to politics. The more context a job requires, the less likely a computer will be able to do it soon. Common sense is important not just because your mom taught you so, but because computers don't have it.

The best way to not lose your job is to automate it yourself. Then you'll have time for all the parts of it that you didn't before and that a computer won't be able to do any time soon. (If there aren't any, stay ahead of the curve and get a new job now.) If a computer has learned to do your job, don't try to compete with it; harness it. H&R Block is still in business, but tax preparers' jobs are much less dreary than they used to be, now that computers do most of the grunge work. (OK, perhaps this is not the best example, given that the tax code's exponential growth is one of the few that can hold its own against computing power's exponential growth.) Think of big data as an extension of your senses and learning algorithms as an extension of your brain. The best chess players these days are so-called centaurs, half-man and half-program. The same is true in many other occupations, from stock analyst to baseball scout. It's not man versus machine; it's man with machine versus man without. Data and intuition are like horse and rider, and you don't try to outrun a horse; you ride it.

As technology progresses, an ever more intimate mix of human and machine takes shape. You're hungry; Yelp suggests some good restaurants. You pick one; GPS gives you directions. You drive; car electronics does the low-level control. We are all cyborgs already. The real story of automation is not what it replaces but what it enables. Some professions disappear, but many more are born. Most of all, automation makes all sorts of things possible that would be way too expensive if done by humans. ATMs replaced some bank tellers, but mainly they let us withdraw money any time, anywhere. If pixels had to be colored one at a time by human animators, there would be no *Toy Story* and no video games.

Still, we can ask whether we'll eventually run out of jobs for humans. I think not. Even if the day comes—and it won't be soon—when computers and robots can do everything better, there will still be jobs for at least some of us. A robot may be able to do a perfect impersonation of a bartender, down to the small talk, but patrons may still prefer a bartender they know is human, just because he is. Restaurants with human waiters will have extra cachet, just as handmade goods already do. People still go to the theater, ride horses, and sail, even though we have movies, cars, and motorboats. More importantly, some professionals will be truly irreplaceable because their jobs require the one thing that computers and robots by definition cannot have: the human experience. By that I don't mean touchy-feely jobs, because touchy-feely is not hard to fake; witness the success of robo-pets. I mean the humanities, whose domain is precisely everything you can't understand without the experience of being human. We worry that the humanities are in a death spiral, but they'll rise from the ashes once other professions have been automated. The more everything is done cheaply by machines, the more valuable the humanist's contribution will be.

Conversely, the long-term prospects of scientists are not the brightest, sadly. In the future, the only scientists may well be computer scientists, meaning computers doing science. The people formerly known as scientists (like me) will devote their lives to understanding the scientific advances made by computers. They won't be noticeably less happy than before; after all, science was always a hobby to them. And one very important job for the technically minded will remain: keeping an eye on the computers. In fact, this will require more than engineers; ultimately, it may be the full-time occupation of all mankind to figure out what we want from the machines and make sure we're getting it—more on this later in this chapter.

In the meantime, as the boundary between automatable and non-automatable jobs advances across the economic landscape, what we'll likely see is unemployment creeping up, downward pressure on the wages of more and more professions, and increasing rewards for the fewer and fewer that can't yet be automated. This is what's already happening, of course, but it has much further to run. The transition will be tumultuous, but thanks to democracy, it will have a happy ending. (Hold on to your vote—it may be the most valuable thing you have.) When the unemployment rate rises above 50 percent, or even before, attitudes about redistribution will radically change. The newly unemployed majority will vote for generous lifetime unemployment benefits and the sky-high taxes needed to fund them. These won't break the bank because machines will do the necessary production. Eventually, we'll start talking about the employment rate instead of the unemployment one and reducing it will be seen as a sign of progress. ("The US is falling behind. Our employment rate is still 23 percent.") Unemployment benefits will be replaced by a basic income for everyone. Those of us who aren't satisfied with it will be able to earn more, stupendously more, in the few remaining human occupations. Liberals and conservatives will still fight about the tax rate, but the goalposts will have permanently moved. With the total value of labor greatly reduced, the wealthiest nations will be those with the highest ratio of natural resources to population. (Move to Canada now.) For those of us not working, life will not be meaningless, any more than life on a tropical island where nature's bounty meets all needs is meaningless. A gift economy will develop, of which the open-source software movement is a preview. People will seek meaning in human relationships, self-actualization, and spirituality, much as they do now. The need to earn a living will be a distant memory, another piece of humanity's barbaric past that we rose above.

War is not for humans

Soldiering is harder to automate than science, but it will be as well. One of the prime uses of robots is to do things that are too dangerous for humans, and fighting wars is about as dangerous as it gets. Robots already defuse bombs, and drones allow a platoon to see over the hill. Self-driving supply trucks and robotic mules are on the way. Soon we will need to decide whether robots are allowed to pull the trigger on their own. The argument for doing this is that we want to get humans out of harm's way, and remote control is not viable in fast-moving, shoot-or-be-shot situations. The argument against is that robots don't understand ethics, and so can't be entrusted with life-or-death decisions. But we can teach them. The deeper question is whether we're ready to.

It's not hard to state general principles like military necessity, proportionality, and sparing civilians. But there's a gulf between them and concrete actions, which the soldier's judgment has to bridge. Asimov's three laws of robotics quickly run into trouble when robots try to apply them in practice, as his stories memorably illustrate. General principles are usually contradictory, if not self-contradictory, and they have to be lest they turn all shades of gray into black and white. When does military necessity outweigh sparing civilians? There is no universal answer and no way to program a computer with all the eventualities. Machine learning, however, provides an alternative. First, teach the robot to recognize the relevant concepts, for example with data sets of situations where civilians were and were not spared, armed response was and was not proportional, and so on. Then give it a code of conduct in the form of rules involving these concepts. Finally, let the robot learn how to apply the code by observing humans: the soldier opened fire in this case but not in that case. By generalizing from these examples, the robot can learn an end-to-end model of ethical decision making, in the form of, say, a large MLN. Once the robot's decisions agree with a human's as often as one human agrees with another, the training is complete, meaning the model is ready for download into thousands of robot brains. Unlike humans, robots don't lose their heads in the heat of combat. If a robot malfunctions, the manufacturer is responsible. If it makes a wrong call, its teachers are.

The main problem with this scenario, as you may have already guessed, is that letting robots learn ethics by observing humans may not be such a good idea. The robot is liable to get seriously confused when it sees that humans' actions often violate their ethical principles. We can clean up the training data by including only the examples where, say, a panel of ethicists agrees that the soldier made the right decision, and the panelists can also inspect and tweak the model postlearning to their satisfaction. Agreement may be hard to reach, however, particularly if the panel includes all the different kinds of people it should. Teaching ethics to robots, with their logical minds and lack of baggage, will force us to examine our assumptions and sort out our contradictions. In this, as in many other areas, the greatest benefit of machine learning may ultimately be not what the machines learn but what we learn by teaching them.

Another objection to robot armies is that they make war too easy. But if we unilaterally relinquish them, that could cost us the next war. The logical response, advocated by the United Nations and Human Rights Watch, is a treaty banning robot warfare, similar to the Geneva Protocol of 1925 banning chemical and biological warfare. This misses a crucial distinction, however. Chemical and biological warfare can only increase human suffering, but robot warfare can greatly decrease it. If a war is fought by machines, with humans only in command positions, no one is killed or wounded. Perhaps, then, what we should do, instead of outlawing robot soldiers, is—when we're ready—outlaw human soldiers.

Robot armies may indeed make wars more likely, but they will also change the ethics of war. Shoot/don't shoot dilemmas become much easier if the targets are other robots. The modern view of war as an unspeakable horror, to be engaged in only as a last resort, will give way to a more nuanced view of war as an orgy of destruction that leaves all sides impoverished and is best avoided but not at all costs. And if war is reduced to a competition to see who can destroy the most, then why not compete instead to create the most?

In any case, banning robot warfare may not be viable. Far from banning drones—the precursors of tomorrow's warbots—countries large and small are busy developing them, presumably because in their estimation the benefits outweigh the risks. As with any weapon, it's safer to have robots than to trust the other side not to. If in future wars millions of kamikaze drones will destroy conventional armies in minutes, they'd better be our drones. If World War III will be over in seconds, as one side takes control of the other's systems, we'd better have the smarter, faster, more resilient network. (Off-grid systems are not the answer: systems that aren't networked can't be hacked, but they can't compete with networked systems, either.) And, on balance, a robot arms race may be a good thing, if it hastens the day when the Fifth Geneva

Convention bans humans in combat. War will always be with us, but the casualties of war need not be.

Google + Master Algorithm = Skynet?

Of course, robot armies also raise a whole different specter. According to Hollywood, the future of humanity is to be snuffed out by a gargantuan AI and its vast army of machine minions. (Unless, of course, a plucky hero saves the day in the last five minutes of the movie.) Google already has the gargantuan hardware such an AI would need, and it's recently acquired an army of robotics startups to go with it. If we drop the Master Algorithm into its servers, is it game over for humanity? Why yes, of course. It's time to reveal my true agenda, with apologies to Tolkien:

> *Three Algorithms for the Scientists under the sky,*
> *Seven for the Engineers in their halls of servers,*
> *Nine for Mortal Businesses doomed to die,*
> *One for the Dark AI on its dark throne,*
> *In the Land of Learning where the Data lies.*
> *One Algorithm to rule them all, One Algorithm to find them,*
> *One Algorithm to bring them all and in the darkness bind them,*
> *In the Land of Learning where the Data lies.*

Hahahaha! Seriously, though, should we worry that machines will take over? The signs seem ominous. With every passing year, computers don't just do more of the world's work; they make more of the decisions. Who gets credit, who buys what, who gets what job and what raise, which stocks will go up and down, how much insurance costs, where police officers patrol and therefore who gets arrested, how long their prison terms will be, who dates whom and therefore who will be born: machine-learned models already play a part in all of these. The point where we could turn off all our computers without causing the collapse of modern civilization has long passed. Machine learning is the last straw: if computers can start programming themselves, all hope of controlling them is surely lost. Distinguished scientists like Stephen Hawking have called for urgent research on this issue before it's too late.

Relax. The chances that an AI equipped with the Master Algorithm will take over the world are *zero*. The reason is simple: unlike humans, computers don't have a will of their own. They're products of engineering, not evolution. Even an infinitely powerful computer would still be only an extension of our will and nothing to fear. Recall the three components of every learning algorithm: representation, evaluation, and optimization. The learner's representation circumscribes what it can learn. Let's make it a very powerful one, like Markov logic, so the learner can in principle learn anything. The optimizer then does everything in its power to maximize the evaluation function—no more and no less—and the evaluation function is *determined by us*. A more powerful computer will just optimize it better. There's no risk of it getting out of control, even if it's a genetic algorithm. A learned system that didn't do what we want would be severely unfit and soon die out. In fact, it's the systems that have even a slight edge in serving us better that will, generation after generation, multiply and take over the gene pool. Of course, if we're so foolish as to deliberately program a computer to put itself above us, then maybe we'll get what we deserve.

The same reasoning applies to all AI systems because they all—explicitly or implicitly—have the same three components. They can vary what they do, even come up with surprising plans, but only in service of the goals we set them. A robot whose programmed goal is "make a good dinner" may decide to cook a steak, a bouillabaisse, or even a delicious new dish of its own creation, but it can't decide to murder its owner any more than a car can decide to fly away. The purpose of AI systems is to solve NP-complete problems, which, as you may recall from Chapter 2, may take exponential time, but the solutions can always be checked efficiently. We should therefore welcome with open arms computers that are vastly more powerful than our brains, safe in the knowledge that our job is exponentially easier than theirs. They have to solve the problems; we just have to check that they did so to our satisfaction. AIs will think fast what we think slow, and the world will be the better for it. I, for one, welcome our new robot underlings.

It's natural to worry about intelligent machines taking over because the only intelligent entities we know are humans and other animals, and they definitely have a will of their own. But there is no necessary connection between intelligence and autonomous will; or rather, intelligence and will may not inhabit the same body, provided there is a line of control between them. In *The Extended Phenotype*, Richard Dawkins shows how nature is replete with examples of an animal's genes controlling more than its own body, from cuckoo eggs to beaver dams. Technology is the extended phenotype of man. This means we can continue to control it even if it becomes far more complex than we can understand.

Picture two strands of DNA going for a swim in their private pool, aka a bacterium's cytoplasm, two billion years ago. They're pondering a momentous decision. "I'm worried, Diana," says one. "If we start making multicellular creatures, will they take over?" Fast-forward to the twenty-first century, and DNA is still alive and well. Better than ever, in fact, with an increasing fraction living safely in bipedal organisms comprising trillions of cells. It's been quite a ride for our tiny double-stranded friends since they made their momentous decision. Humans are their trickiest creation yet; we've invented things like contraception that let us have fun without spreading our DNA, and we have—or seem to have—free will. But it's still DNA that shapes our notions of fun, and we use our free will to pursue pleasure and avoid pain, which, for the most part, still coincides with what's best for our DNA's survival. We may yet be DNA's demise if we choose to transmute ourselves into silicon, but even then, it's been a great two billion years. The decision we face today is similar: if we start making AIs—vast, interconnected, superhuman, unfathomable AIs—will they take over? Not any more than multicellular organisms took over from genes, vast and unfathomable as we may be to them. AIs are our survival machines, in the same way that we are our genes'.

This does not mean that there is nothing to worry about, however. The first big worry, as with any technology, is that AI could fall into the wrong hands. If a criminal or prankster programs an AI to take over the world, we'd better have an AI police capable of catching it and erasing it before it gets too far. The best insurance policy against vast AIs gone amok is vaster AIs keeping the peace.

The second worry is that humans will voluntarily surrender control. It starts with robot rights, which seem absurd to me but not to everyone. After all, we already give rights to animals, who never asked for them. Robot rights might seem like the logical next step in expanding the "circle of empathy." Feeling empathy for robots is not hard, particularly if they're designed to elicit it. Even Tamagotchi, Japanese "virtual pets" with all of three buttons and an LCD screen, do it quite successfully. The first humanoid consumer robot will set off a race to make more and more empathy-eliciting robots, because they'll sell much better than the plain metal variety. Children raised by robot nannies will have a life-long soft spot for kindly electronic friends. The "uncanny valley"—our discomfort with robots that are almost human but not quite—will be unknown to them because they grew up with robot mannerisms and maybe even adopted them as cool teenagers.

The next step in the insidious progression of AI control is letting them make all the decisions because they're, well, so much smarter. Beware. They may be smarter, but they're in the service of whoever designed their score functions. This is the "Wizard of Oz" problem. Your job in a world of intelligent machines is to keep making sure they do what you want, both at the input (setting the goals) and at the output (checking that you got what you asked for). If you don't, somebody else will. Machines can help us figure out collectively what we want, but if you don't participate, you lose out—just like democracy, only more so. Contrary to what we like to believe today, humans quite easily fall into obeying others, and any sufficiently advanced AI is indistinguishable from God. People won't necessarily mind taking their marching orders from some vast oracular computer; the question is who oversees the overseer. Is AI the road to a more perfect democracy or to a more insidious dictatorship? The eternal vigil has just begun.

The third and perhaps biggest worry is that, like the proverbial genie, the machines will give us what we ask for instead of what we want. This is not a hypothetical scenario; learning algorithms do it all the time. We train a neural network to recognize horses, but it learns instead to recognize brown patches, because all the horses in its training set happened to be brown. You just bought a watch, so Amazon recommends similar items: other watches, which are now the last thing you want to buy. If you examine all the decisions that computers make today—who gets credit, for example—you'll find that they're often needlessly bad. Yours would be too, if your brain was a support vector machine and all your knowledge of credit scoring came from perusing one lousy database. People worry that computers will get too smart and take over the world, but the real problem is that they're too stupid and they've already taken over the world.

Evolution, part 2

Even if computers today are still not terribly smart, there's no doubt that their intelligence is rapidly increasing. As early as 1965, I. J. Good, a British statistician and Alan Turing's sidekick on the World War II Enigma code-breaking project, speculated on a coming intelligence explosion. Good pointed out that if we can design machines that are more intelligent than us, they should in turn be able to design machines that are more intelligent than them, and so on ad infinitum, leaving human intelligence far behind. In a 1993 essay, Vernor Vinge christened this "the Singularity." The concept has been popularized most of all by Ray Kurzweil, who argues in *The Singularity Is Near* that not only is the Singularity inevitable, but the point where machine intelligence exceeds human intelligence—let's call it the Turing point—will arrive within the next few decades.

Clearly, without machine learning—programs that design programs—the Singularity cannot happen. We also need sufficiently powerful hardware, but that's coming along nicely. We'll reach the Turing point soon after we invent the Master Algorithm. (I'm willing to bet Kurzweil a bottle of Dom Pérignon that this will happen before we reverse engineer the brain, his method of choice for bringing about human-level AI.) *Pace* Kurzweil, this will not, however, lead to the Singularity. It will lead to something much more interesting.

The term *singularity* comes from mathematics, where it denotes a point at which a function becomes infinite. For example, the function $1/x$ has a singularity when x is 0, because 1 divided by 0 is infinity. In physics, the quintessential example of a singularity is a black hole: a point of infinite density, where a finite amount of matter is crammed into infinitesimal space. The only problem with singularities is that they don't really exist. (When did you last divide a cake among zero people, and each one got an infinite slice?) In physics, if a theory predicts something is infinite, something's wrong with the theory. Case in point, general relativity presumably predicts that black holes have infinite density because it ignores quantum effects. Likewise, intelligence cannot continue to increase forever. Kurzweil acknowledges this, but points to a series of exponential curves in technology improvement (processor speed, memory capacity, etc.) and argues that the limits to this growth are so far away that we need not concern ourselves with them.

Kurzweil is overfitting. He correctly faults other people for always extrapolating linearly—seeing straight lines instead of curves—but then falls prey to a more exotic malady: seeing exponentials everywhere. In curves that are flat—nothing happening—he sees exponentials that have not taken off yet. But technology improvement curves are not exponentials; they are S curves, our good friends from Chapter 4. The early part of an S curve is easy to mistake for an exponential, but then they quickly diverge. Most of Kurzweil's curves are consequences of Moore's law, which is on its last legs. Kurzweil argues that other technologies will take the place of semiconductors and S curve will pile on S curve, each steeper than the previous one, but this is speculation. He goes even further to claim that the entire history of life on Earth, not just human technology, shows exponentially accelerating progress, but this perception is at least partly due to a parallax effect: things that are closer seem to move faster. Trilobites in the heat of the Cambrian explosion could be forgiven for believing in exponentially accelerating progress, but then there was a big slowdown. A Tyrannosaurus Ray would probably have proposed a law of accelerating body size. Eukaryotes (us) evolve more slowly than prokaryotes (bacteria). Far from accelerating smoothly, evolution proceeds in fits and starts.

To sidestep the problem that infinitely dense points don't exist, Kurzweil proposes to instead equate the Singularity with a black hole's event horizon, the region within which gravity is so strong that not even light can escape. Similarly, he says, the Singularity is the point beyond which technological evolution is so fast that humans cannot predict or understand what will happen. If that's what the Singularity is, then we're already inside it. We can't predict in advance what a learner will come up with, and often we can't even understand it in retrospect. As a matter of fact, we've always lived in a world that we only partly understood. The main difference is that our world is now partly created by us, which is surely an improvement. The world beyond the Turing point will not be incomprehensible to us, any more than the Pleistocene was. We'll focus on what we can understand, as we always have, and call the rest random (or divine).

The trajectory we're on is not a singularity but a phase transition. Its critical point—the Turing point—will come when machine learning overtakes the natural variety. Natural learning itself has gone through three phases: evolution, the brain, and culture. Each is a product of the previous one, and each learns faster. Machine learning is the logical next stage of this progression. Computer programs are the fastest replicators on Earth: copying them takes only a fraction of a second. But creating them is slow, if it has to be done by humans. Machine learning removes that bottleneck, leaving a final one: the speed at which humans can absorb change. This too will eventually be removed, but not because we'll decide to hand things off to our "mind children," as Hans Moravec calls them, and go gently into the good night. Humans are not a dying twig on the tree of life. On the contrary, we're about to start branching.

In the same way that culture coevolved with larger brains, we will coevolve with our creations. We always have: humans would be physically different if we had not invented fire or spears. We are *Homo technicus* as much as *Homo sapiens*. But a model of the cell of the kind I envisaged in the last chapter will allow something entirely new: computers that design cells based on the parameters we give them, in the same way that silicon compilers design microchips based on their functional specifications. The corresponding DNA can then be synthesized and inserted into a "generic" cell, transforming it into the desired one. Craig Venter, the genome pioneer, has already taken the first steps in this direction. At first we will use this power to fight disease: a new pathogen is identified, the cure is immediately found, and your immune system downloads it from the Internet. *Health problems* becomes an oxymoron. Then DNA design will let people at last have the body they want, ushering in an age of affordable beauty, in William Gibson's memorable words. And then *Homo technicus* will evolve into a myriad different intelligent species, each with its own niche, a whole new biosphere as different from today's as today's is from the primordial ocean.

Many people worry that human-directed evolution will permanently split the human race into a class of genetic haves and one of have-nots. This strikes me as a singular failure of imagination. Natural evolution did not result in just two species, one subservient to the other, but in an infinite variety of creatures and intricate ecosystems. Why would artificial evolution, building on it but less constrained, do so?

Like all phase transitions, this one will eventually taper off too. Overcoming a bottleneck does not mean the sky is the limit; it means the next bottleneck is the limit, even if we don't see it yet. Other transitions will follow, some large, some small, some soon, some not for a long time. But the next thousand years could well be the most amazing in the life of planet Earth.

Epilogue

So now you know the secrets of machine learning. The engine that turns data into knowledge is no longer a black box: you know how the magic happens and what it can and can't do. You've met the complexity monster, the overfitting problem, the curse of dimensionality, and the exploration-exploitation dilemma. You know in broad outline what Google, Facebook, Amazon, and all the rest do with the data you generously give them every day and why they can find stuff for you, filter out spam, and keep improving their offerings. You've seen what's brewing in the world's machine-learning research labs, and you have a ringside seat to the future they're helping to bring about. You've met the five tribes of machine learning and their master algorithms: symbolists and inverse deduction; connectionists and backpropagation; evolutionaries and genetic algorithms; Bayesians and probabilistic inference; analogizers and support vector machines. And because you've traveled over a vast territory, negotiated the border crossings, and climbed the high peaks, you have a better view of the landscape than even many machine learners, who toil daily in the fields. You can see the common themes running through the land like an underground river, and you know how the five master algorithms, superficially so different, are really just five facets of a single universal learner.

But the journey is far from over. We don't have the Master Algorithm yet, just a glimpse of what it might look like. What if something fundamental is still missing, something all of us in the field, steeped in its history, can't see? We need new ideas, and ideas that are not just variations on the ones we already have. That's why I wrote this book: to start you thinking. I teach an evening class on machine learning at the University of Washington. In 2007, soon after the Netflix Prize was announced, I proposed it as one of the class projects. Jeff Howbert, a student in the class, got hooked and continued to work on it after the class was over. He wound up being a member of one of the two winning teams, two years after learning about machine learning for the first time. Now it's your turn. To learn more about machine learning, check out the section on further readings at the end of the book. Download some data sets from the UCI repository (archive.ics.uci.edu/ml/) and start playing. When you're ready, check out Kaggle.com, a whole website dedicated to running machine-learning competitions, and pick one or two to enter. Of course, it'll be more fun if you recruit a friend or two to work with you. If you're hooked, like Jeff was, and wind up becoming a professional data scientist, welcome to the most fascinating job in the world. If you find yourself dissatisfied with today's learners, invent new ones—or just do it for fun. My fondest wish is that your reaction to this book will be like my reaction to the first AI book I read, over twenty years ago: there's so much to do here, I don't know where to start. If one day you invent the Master Algorithm, please don't run to the patent office with it. Open-source it. The Master Algorithm is too important to be owned by any one person or organization. Its applications will multiply faster than you can license it. But if you decide instead to do a startup, remember to give a share in it to every man, woman, and child on Earth.

Whether you read this book out of curiosity or professional interest, I hope you will share what you've learned with your friends and colleagues. Machine learning touches the lives of every one of us, and it's up to all of us to decide what we want to do with it. Armed with your new understanding of machine learning, you're in a much better position to think about issues like privacy and data sharing, the future of work, robot warfare, and the promise and peril of AI; and the more of us have this understanding, the more likely we'll avoid the pitfalls and find the right paths. That's the other big reason I wrote this book. The statistician knows that prediction is hard, especially about the future, and the computer scientist knows that the best way to predict the future is to invent it, but the unexamined future is not worth inventing.

Thanks for letting me be your guide. I'd like to give you a parting gift. Newton said that he felt like a boy playing on the seashore, picking up a pebble here and a shell there while the great ocean of truth lay undiscovered before him. Three hundred years later, we've gathered an amazing collection of pebbles and shells, but the great undiscovered ocean still stretches into the distance, sparkling with promise. The gift is a boat—machine learning—and it's time to set sail.

Acknowledgments

First of all, I thank my companions in scientific adventure: students, collaborators, colleagues, and everyone in the machine-learning community. This is your book as much as mine. I hope you will forgive my many oversimplifications and omissions, and the somewhat fanciful way in which parts of the book are written.

I'm grateful to everyone who read and commented on drafts of the book at various stages, including Mike Belfiore, Thomas Dietterich, Tiago Domingos, Oren Etzioni, Abe Friesen, Rob Gens, Alon Halevy, David Israel, Henry Kautz, Chloé Kiddon, Gary Marcus, Ray Mooney, Kevin Murphy, Franzi Roesner, and Ben Taskar. Thanks also to everyone who gave me pointers, information, or help of various kinds, including Tom Griffiths, David Heckerman, Hannah Hickey, Albert-László Barabási, Yann LeCun, Barbara Mones, Mike Morgan, Peter Norvig, Judea Pearl, Gregory Piatetsky-Shapiro, and Sebastian Seung.

I'm lucky to work in a very special place, the University of Washington's Department of Computer Science and Engineering. I'm also grateful to Josh Tenenbaum, and to everyone in his group, for hosting the sabbatical at MIT during which I started this book. Thanks to Jim Levine, my indefatigable agent, for drinking the Kool-Aid (as he put it) and spreading the word; and to everyone at Levine Greenberg Rostan. Thanks to TJ Kelleher, my amazing editor, for helping make this a better book, chapter by chapter, line by line; and to everyone at Basic Books.

I'm indebted to the organizations that have funded my research over the years, including ARO, DARPA, FCT, NSF, ONR, Ford, Google, IBM, Kodak, Yahoo, and the Sloan Foundation.

Last and most, I thank my family for their love and support.

Further Readings

If this book whetted your appetite for machine learning and the issues surrounding it, you'll find many suggestions in this section. Its aim is not to be comprehensive but to provide an entrance to machine learning's garden of forking paths (as Borges put it). Wherever possible, I chose books and articles appropriate for the general reader. Technical publications, which require at least some computational, statistical, or mathematical background, are marked with an asterisk (*). Even these, however, often have large sections accessible to the general reader. I didn't list volume, issue, or page numbers, since the web renders them superfluous; likewise for publishers' locations.

If you'd like to learn more about machine learning in general, one good place to start is online courses. Of these, the closest in content to this book is, not coincidentally, the one I teach (http://www.cs.washington.edu/homes/pedrod/class). Two other options are Andrew Ng's course (www.coursera.org/course/ml) and Yaser Abu-Mostafa's (http://work.caltech.edu/telecourse.html). The next step is to read a textbook. The closest to this book, and one of the most accessible, is Tom Mitchell's *Machine Learning** (McGraw-Hill, 1997). More up-to-date, but also more mathematical, are Kevin Murphy's *Machine Learning: A Probabilistic Perspective** (MIT Press, 2012), Chris Bishop's *Pattern Recognition and Machine Learning** (Springer, 2006), and *An Introduction to Statistical Learning with Applications in R,** by Gareth James, Daniela Witten, Trevor Hastie, and Rob Tibshirani (Springer, 2013). My article "A few useful things to know about machine learning" (*Communications of the ACM*, 2012) summarizes some of the "folk knowledge" of machine learning that textbooks often leave implicit and was one of the starting points for this book. If you know how to program and are itching to give machine learning a try, you can start from a number of open-source packages, such as Weka (www.cs.waikato. ac.nz/ml/weka). The two main machine-learning journals are *Machine Learning* and the *Journal of Machine Learning Research.* Leading machine-learning conferences, with yearly proceedings, include the International Conference on Machine Learning, the Conference on Neural Information Processing Systems, and the International Conference on Knowledge Discovery and Data Mining. A large number of machine-learning talks are available on http://videolectures.net. The www. KDnuggets.com website is a one-stop shop for machine-learning resources, and you can sign up for its newsletter to keep up-to-date with the latest developments.

Prologue

An early list of examples of machine learning's impact on daily life can be found in "Behind-the-scenes data mining," by George John (*SIGKDD Explorations,* 1999), which was also the inspiration for the "day-in-the-life" paragraphs of the prologue. Eric Siegel's book *Predictive Analytics* (Wiley, 2013) surveys a large number of machine-learning applications. The term *big data* was popularized by the McKinsey Global Institute's 2011 report *Big Data: The Next Frontier for Innovation, Competition, and Productivity.* Many of the issues raised by big data are discussed in *Big Data: A Revolution That Will Change How We Live, Work, and Think,* by Viktor Mayer-Schönberger and Kenneth Cukier (Houghton Mifflin Harcourt, 2013). The textbook I learned AI from is *Artificial Intelligence,** by Elaine Rich (McGraw-Hill, 1983). A current one is *Artificial Intelligence: A Modern Approach,* by Stuart Russell and Peter Norvig (3rd ed., Prentice Hall, 2010). Nils Nilsson's *The Quest for Artificial Intelligence* (Cambridge University Press, 2010) tells the story of AI from its earliest days.

Chapter One

Nine Algorithms That Changed the Future, by John MacCormick (Princeton University Press, 2012), describes some of the most important algorithms in computer science, with a chapter on machine learning. *Algorithms,** by Sanjoy Dasgupta, Christos Papadimitriou, and Umesh Vazirani (McGraw-Hill, 2008), is a concise introductory textbook on the subject. *The Pattern on the Stone,* by Danny Hillis (Basic Books, 1998), explains how computers work. Walter Isaacson recounts the lively history of computer science in *The Innovators* (Simon & Schuster, 2014).

"Spreadsheet data manipulation using examples,"* by Sumit Gulwani, William Harris, and Rishabh Singh (*Communications of the ACM,* 2012), is an example of how computers can program themselves by observing users. *Competing on Analytics*, by Tom Davenport and Jeanne Harris (HBS Press, 2007), is an introduction to the use of predictive analytics in business. *In the Plex*, by Steven Levy (Simon & Schuster, 2011), describes at a high level how Google's technology works. Carl Shapiro and Hal Varian explain the network effect in *Information Rules* (HBS Press, 1999). Chris Anderson does the same for the long-tail phenomenon in *The Long Tail* (Hyperion, 2006).

The transformation of science by data-intensive computing is surveyed in *The Fourth Paradigm*, edited by Tony Hey, Stewart Tansley, and Kristin Tolle (Microsoft Research, 2009). "Machine science," by James Evans and Andrey Rzhetsky (*Science*, 2010), discusses some of the different ways computers can make scientific discoveries. *Scientific Discovery: Computational Explorations of the Creative Processes*,* by Pat Langley et al. (MIT Press, 1987), describes a series of approaches to automating the discovery of scientific laws. The SKICAT project is described in "From digitized images to online catalogs," by Usama Fayyad, George Djorgovski, and Nicholas Weir (*AI Magazine*, 1996). "Machine learning in drug discovery and development,"* by Niki Wale (*Drug Development Research*, 2001), gives an overview of just that. Adam, the robot scientist, is described in "The automation of science," by Ross King et al. (*Science*, 2009).

Sasha Issenberg's *The Victory Lab* (Broadway Books, 2012) dissects the use of data analysis in politics. "How President Obama's campaign used big data to rally individual votes," by the same author (*MIT Technology Review*, 2013), tells the story of its greatest success to date. Nate Silver's *The Signal and the Noise* (Penguin Press, 2012) has a chapter on his poll aggregation method.

Robot warfare is the theme of P. W. Singer's *Wired for War* (Penguin, 2009). *Cyber War*, by Richard Clarke and Robert Knake (Ecco, 2012), sounds the alarm on cyberwar. My work on combining machine learning with game theory to defeat adversaries, which started as a class project, is described in "Adversarial classification,"* by Nilesh Dalvi et al. (*Proceedings of the Tenth International Conference on Knowledge Discovery and Data Mining*, 2004). *Predictive Policing*, by Walter Perry et al. (Rand, 2013), is a guide to the use of analytics in police work.

Chapter Two

The ferret brain rewiring experiments are described in "Visual behaviour mediated by retinal projections directed to the auditory pathway," by Laurie von Melchner, Sarah Pallas, and Mriganka Sur (*Nature*, 2000). Ben Underwood's story is told in "Seeing with sound," by Joanna Moorhead (*Guardian*, 2007), and at www.ben underwood.com. Otto Creutzfeldt makes the case that the cortex is one algorithm in "Generality of the functional structure of the neocortex" (*Naturwissenschaften*,

1977), as does Vernon Mountcastle in "An organizing principle for cerebral function: The unit model and the distributed system," in *The Mindful Brain*, edited by Gerald Edelman and Vernon Mountcastle (MIT Press, 1978). Gary Marcus, Adam Marblestone, and Tom Dean make the case against in "The atoms of neural computation" (*Science*, 2014).

"The unreasonable effectiveness of data," by Alon Halevy, Peter Norvig, and Fernando Pereira (*IEEE Intelligent Systems*, 2009), argues for machine learning as the new discovery paradigm. Benoît Mandelbrot explores the fractal geometry of nature in the eponymous book* (Freeman, 1982). James Gleick's *Chaos* (Viking, 1987) discusses and depicts the Mandelbrot set. The Langlands program, a research effort that seeks to unify different subfields of mathematics, is described in *Love and Math*, by Edward Frenkel (Basic Books, 2014). *The Golden Ticket*, by Lance Fortnow (Princeton University Press, 2013), is an introduction to NP-completeness and the P = NP problem. *The Annotated Turing*,* by Charles Petzold (Wiley, 2008), explains Turing machines by revisiting Turing's original paper on them.

The Cyc project is described in "Cyc: Toward programs with common sense,"* by Douglas Lenat et al. (*Communications of the ACM*, 1990). Peter Norvig discusses Noam Chomsky's criticisms of statistical learning in "On Chomsky and the two cultures of statistical learning" (http://norvig.com/chomsky.html). Jerry Fodor's *The Modularity of Mind* (MIT Press, 1983) summarizes his views on how the mind works. "What big data will never explain," by Leon Wieseltier (*New Republic*, 2013), and "Pundits, stop sounding ignorant about data," by Andrew McAfee (*Harvard Business Review*, 2013), give a flavor of the controversy surrounding what big data can and can't do. Daniel Kahneman explains why algorithms often beat intuitions in Chapter 21 of *Thinking, Fast and Slow* (Farrar, Straus and Giroux, 2011). David Patterson makes the case for the role of computing and data in the fight against cancer in "Computer scientists may have what it takes to help cure cancer" (*New York Times*, 2011).

More on the various tribes' paths to the Master Algorithm in the corresponding sections below.

Chapter Three

Hume's classic formulation of the problem of induction appears in Volume I of *A Treatise of Human Nature* (1739). David Wolpert derives his "no free lunch" theorem for induction in "The lack of a priori distinctions between learning algorithms"* (*Neural Computation*, 1996). I discuss the importance of prior knowledge in machine learning in "Toward knowledge-rich data mining"* (*Data Mining and Knowledge Discovery*, 2007) and misinterpretations of Occam's razor in "The role of Occam's razor in knowledge discovery"* (*Data Mining and Knowledge Discovery*, 1999). Overfitting is one of the main themes of *The Signal and the Noise*, by Nate Silver (Penguin Press, 2012), who calls it "the most important scientific problem you've never heard of." "Why most published research findings are false,"* by John Ioannidis (*PLoS Medicine*, 2005), discusses the problem of mistaking chance findings for true ones in science. Yoav Benjamini and Yosef Hochberg propose a way to combat it in "Controlling the false discovery rate: A practical and powerful approach to multiple testing"* (*Journal of the Royal Statistical Society, Series B*, 1995). The bias-variance decomposition is presented in "Neural networks and the bias/variance dilemma," by Stuart Geman, Elie Bienenstock, and René Doursat (*Neural Computation*, 1992). "Machine learning as an experimental science," by Pat Langley (*Machine Learning*, 1988), discusses the role of experimentation in machine learning.

William Stanley Jevons first proposed viewing induction as the inverse of deduction in *The Principles of Science* (1874). The paper "Machine learning of first-order predicates by inverting resolution,"* by Steve Muggleton and Wray Buntine (*Proceedings of the Fifth International Conference on Machine Learning*, 1988), initiated the use of inverse deduction in machine learning. The book *Relational Data Mining*,* edited by Sašo Džeroski and Nada Lavrač (Springer, 2001), is an introduction to the field of inductive logic programming, where inverse deduction is studied. "The CN2 Induction Algorithm,"* by Peter Clark and Tim Niblett (*Machine Learning*, 1989), summarizes some of the main Michalski-style rule induction algorithms. The rule-mining approach used by retailers is described in "Fast algorithms for mining association rules,"* by Rakesh Agrawal and Ramakrishnan Srikant (*Proceedings of the Twentieth International Conference on Very Large Databases*, 1994). An example of rule induction for cancer prediction is described in "Carcinogenesis predictions using inductive logic programming," by Ashwin Srinivasan, Ross King, Stephen Muggleton, and Michael Sternberg (*Intelligent Data Analysis in Medicine and Pharmacology*, 1997).

The two leading decision tree learners are presented in *C4.5: Programs for Machine Learning*,* by J. Ross Quinlan (Morgan Kaufmann, 1992), and *Classification and Regression Trees*,* by Leo Breiman, Jerome Friedman, Richard Olshen, and Charles Stone (Chapman and Hall, 1984). "Real-time human pose recognition in parts from single depth images,"* by Jamie Shotton et al. (*Communications of the ACM*, 2013), explains how Microsoft's Kinect uses decision trees to track gamers' motions. "Competing approaches to predicting Supreme Court decision making," by Andrew Martin et al. (*Perspectives on Politics*, 2004), describes how decision trees beat legal experts at predicting Supreme Court votes and shows the decision tree for Justice Sandra Day O'Connor.

Allen Newell and Herbert Simon formulated the hypothesis that all intelligence is symbol manipulation in "Computer science as empirical enquiry: Symbols and search" (*Communications of the ACM*, 1976). David Marr proposed his three levels of information processing in *Vision** (Freeman, 1982). *Machine Learning: An Artificial Intelligence Approach*,* edited by Ryszard Michalski, Jaime Carbonell, and Tom Mitchell (Tioga, 1983), gives a snapshot of the early days of symbolist research in machine learning. "Connectionist AI, symbolic AI, and the brain,"* by Paul Smolensky (*Artificial Intelligence Review*, 1987), gives a connectionist view of symbolist models.

Chapter Four

Sebastian Seung's *Connectome* (Houghton Mifflin Harcourt, 2012) is an accessible introduction to neuroscience, connectomics, and the daunting challenge of reverse engineering the brain. *Parallel Distributed Processing*,* edited by David Rumelhart, James McClelland, and the PDP research group (MIT Press, 1986), is the bible of connectionism in its 1980s heyday. *Neurocomputing*,* edited by James Anderson and Edward Rosenfeld (MIT Press, 1988), collates many of the classic connectionist papers, including: McCulloch and Pitts on the first models of neurons; Hebb on Hebb's rule; Rosenblatt on perceptrons; Hopfield on Hopfield networks; Ackley, Hinton, and Sejnowski on Boltzmann machines; Sejnowski and Rosenberg on NETtalk; and Rumelhart, Hinton, and Williams on backpropagation. "Efficient backprop,"* by Yann LeCun, Léon Bottou, Genevieve Orr, and Klaus-Robert Müller, in *Neural Networks: Tricks of the Trade*, edited by Genevieve Orr and Klaus-Robert Müller (Springer, 1998), explains some of the main tricks needed to make backprop work.

Neural Networks in Finance and Investing,* edited by Robert Trippi and Efraim Turban (McGraw-Hill, 1992), is a collection of articles on financial applications of neural networks. "Life in the fast lane: The evolution of an adaptive vehicle control system," by Todd Jochem and Dean Pomerleau (*AI Magazine*, 1996), describes the ALVINN self-driving car project. Paul Werbos's PhD thesis is *Beyond Regression: New Tools for Prediction and Analysis in the Behavioral Sciences** (Harvard University, 1974). Arthur Bryson and Yu-Chi Ho describe their early version of backprop in *Applied Optimal Control** (Blaisdell, 1969).

Learning Deep Architectures for AI,* by Yoshua Bengio (Now, 2009), is a brief introduction to deep learning. The problem of error signal diffusion in backprop is described in "Learning long-term dependencies with gradient descent is difficult,"* by Yoshua Bengio, Patrice Simard, and Paolo Frasconi (*IEEE Transactions on Neural Networks*, 1994). "How many computers to identify a cat? 16,000," by

John Markoff (*New York Times*, 2012), reports on the Google Brain project and its results. Convolutional neural networks, the current deep learning champion, are described in "Gradient-based learning applied to document recognition,"* by Yann LeCun, Léon Bottou, Yoshua Bengio, and Patrick Haffner (*Proceedings of the IEEE*, 1998). "The $1.3B quest to build a supercomputer replica of a human brain," by Jonathon Keats (*Wired*, 2013), describes the European Union's brain modeling project. "The NIH BRAIN Initiative," by Thomas Insel, Story Landis, and Francis Collins (*Science*, 2013), describes the BRAIN initiative.

Steven Pinker summarizes the symbolists' criticisms of connectionist models in Chapter 2 of *How the Mind Works* (Norton, 1997). Seymour Papert gives his take on the debate in "One AI or Many?" (*Daedalus*, 1988). *The Birth of the Mind*, by Gary Marcus (Basic Books, 2004), explains how evolution could give rise to the human brain's complex abilities.

Chapter Five

"Evolutionary robotics," by Josh Bongard (*Communications of the ACM*, 2013), surveys the work of Hod Lipson and others on evolving robots. *Artificial Life*, by Steven Levy (Vintage, 1993), gives a tour of the digital zoo, from computer-created animals in virtual worlds to genetic algorithms. Chapter 5 of *Complexity*, by Mitch Waldrop (Touchstone, 1992), tells the story of John Holland and the first few decades of research on genetic algorithms. *Genetic Algorithms in Search, Optimization, and Machine Learning*,* by David Goldberg (Addison-Wesley, 1989), is the standard introduction to genetic algorithms.

Niles Eldredge and Stephen Jay Gould propose their theory of punctuated equilibria in "Punctuated equilibria: An alternative to phyletic gradualism," in *Models in Paleobiology*, edited by T. J. M. Schopf (Freeman, 1972). Richard Dawkins critiques it in Chapter 9 of *The Blind Watchmaker* (Norton, 1986). The exploration-exploitation dilemma is discussed in Chapter 2 of *Reinforcement Learning*,* by Richard Sutton and Andrew Barto (MIT Press, 1998). John Holland proposes his solution, and much else, in *Adaptation in Natural and Artificial Systems** (University of Michigan Press, 1975).

John Koza's *Genetic Programming** (MIT Press, 1992) is the key reference on this paradigm. An evolved robot soccer team is described in "Evolving team *Darwin United*,"* by David Andre and Astro Teller, in *RoboCup-98: Robot Soccer World Cup II*, edited by Minoru Asada and Hiroaki Kitano (Springer, 1999). *Genetic Programming III*,* by John Koza, Forrest Bennett III, David Andre, and Martin Keane (Morgan Kaufmann, 1999), includes many examples of evolved electronic circuits. Danny Hillis argues that parasites are good for evolution in "Co-evolving parasites improve simulated evolution as an optimization procedure"* (*Physica D*, 1990). Adi Livnat, Christos Papadimitriou, Jonathan Dushoff, and Marcus Feldman propose that sex optimizes mixability in "A mixability theory of the role of sex in evolution"* (*Proceedings of the National Academy of Sciences*, 2008). Kevin Lang's paper comparing genetic programming and hill climbing is "Hill climbing beats genetic search on a Boolean circuit synthesis problem of Koza's"* (*Proceedings of the Twelfth International Conference on Machine Learning*, 1995). Koza's reply is "A response to the ML-95 paper entitled . . . "* (unpublished; online at www.genetic-programming.com/jktahoe24page.html).

James Baldwin proposed the eponymous effect in "A new factor in evolution" (*American Naturalist*, 1896). Geoff Hinton and Steven Nowlan describe their implementation of it in "How learning can guide evolution"* (*Complex Systems*, 1987). The Baldwin effect was the theme of a 1996 special issue* of the journal *Evolutionary Computation* edited by Peter Turney, Darrell Whitley, and Russell Anderson.

The distinction between descriptive and normative theories was articulated by John Neville Keynes in *The Scope and Method of Political Economy* (Macmillan, 1891).

Chapter Six

Sharon Bertsch McGrayne tells the history of Bayesianism, from Bayes and Laplace to the present, in *The Theory That Would Not Die* (Yale University Press, 2011). *A First Course in Bayesian Statistical Methods*,* by Peter Hoff (Springer, 2009), is an introduction to Bayesian statistics.

The Naïve Bayes algorithm is first mentioned in *Pattern Classification and Scene Analysis*,* by Richard Duda and Peter Hart (Wiley, 1973). Milton Friedman argues for oversimplified theories in "The methodology of positive economics," which appears in *Essays in Positive Economics* (University of Chicago Press, 1966). The use of Naïve Bayes in spam filtering is described in "Stopping spam," by Joshua Goodman, David Heckerman, and Robert Rounthwaite (*Scientific American*, 2005). "Relevance weighting of search terms,"* by Stephen Robertson and Karen Sparck Jones (*Journal of the American Society for Information Science*, 1976), explains the use of Naïve Bayes–like methods in information retrieval.

"First links in the Markov chain," by Brian Hayes (*American Scientist*, 2013), recounts Markov's invention of the eponymous chains. "Large language models in machine translation,"* by Thorsten Brants et al. (*Proceedings of the 2007 Joint Conference on Empirical Methods in Natural Language Processing and Computational Natural Language Learning*, 2007), explains how Google Translate works.

"The PageRank citation ranking: Bringing order to the Web,"* by Larry Page, Sergey Brin, Rajeev Motwani, and Terry Winograd (Stanford University technical report, 1998), describes the PageRank algorithm and its interpretation as a random walk over the web. *Statistical Language Learning*,* by Eugene Charniak (MIT Press, 1996), explains how hidden Markov models work. *Statistical Methods for Speech Recognition*,* by Fred Jelinek (MIT Press, 1997), describes their application to speech recognition. The story of HMM-style inference in communication is told in "The Viterbi algorithm: A personal history," by David Forney (unpublished; online at arxiv.org/pdf/cs/0504020v2.pdf). *Bioinformatics: The Machine Learning Approach*,* by Pierre Baldi and Søren Brunak (2nd ed., MIT Press, 2001), is an introduction to the use of machine learning in biology, including HMMs. "Engineers look to Kalman filtering for guidance," by Barry Cipra (*SIAM News*, 1993), is a brief introduction to Kalman filters, their history, and their applications.

Judea Pearl's pioneering work on Bayesian networks appears in his book *Probabilistic Reasoning in Intelligent Systems** (Morgan Kaufmann, 1988). "Bayesian networks without tears,"* by Eugene Charniak (*AI Magazine*, 1991), is a largely nonmathematical introduction to them. "Probabilistic interpretation for MYCIN's certainty factors,"* by David Heckerman (*Proceedings of the Second Conference on Uncertainty in Artificial Intelligence*, 1986), explains when sets of rules with confidence estimates are and aren't a reasonable approximation to Bayesian networks. "Module networks: Identifying regulatory modules and their condition-specific regulators from gene expression data," by Eran Segal et al. (*Nature Genetics*, 2003), is an example of using Bayesian networks to model gene regulation. "Microsoft virus fighter: Spam may be more difficult to stop than HIV," by Ben Paynter (*Fast Company*, 2012), tells how David Heckerman took inspiration from spam filters and used Bayesian networks to design a potential AIDS vaccine. The probabilistic or "noisy" OR is explained in Pearl's book.* "Probabilistic diagnosis using a reformulation of the INTERNIST-1/QMR knowledge base," by M. A. Shwe et al. (Parts I and II, *Methods of Information in Medicine*, 1991), describes a noisy-OR Bayesian network for medical diagnosis. Google's Bayesian network for ad placement is described in Section 26.5.4 of Kevin Murphy's *Machine Learning** (MIT Press, 2012). Microsoft's player rating system is described in "TrueSkill™: A Bayesian skill rating system,"* by Ralf Herbrich, Tom Minka, and Thore Graepel (*Advances in Neural Information Processing Systems 19*, 2007).

Modeling and Reasoning with Bayesian Networks,* by Adnan Darwiche (Cambridge University Press, 2009), explains the main algorithms for inference in Bayesian networks. The January/February 2000 issue* of *Computing in Science and Engineering*, edited by Jack Dongarra and Francis Sullivan, has articles on the top ten algorithms of the twentieth century, including MCMC. "Stanley: The robot that won the DARPA Grand Challenge," by Sebastian Thrun et al. (*Journal of Field Robotics*, 2006), explains how the eponymous self-driving car works. "Bayesian networks for data mining,"* by David Heckerman (*Data Mining and Knowledge Discovery*, 1997), summarizes the Bayesian approach to learning and explains how to learn Bayesian networks from data. "Gaussian processes: A replacement for supervised neural networks?,"* by David MacKay (NIPS tutorial notes, 1997; online at www.inference.eng.cam.ac.uk/mackay/gp.pdf), gives a flavor of how the Bayesians co-opted NIPS.

The need for weighting the word probabilities in speech recognition is discussed in Section 9.6 of *Speech and Language Processing*,* by Dan Jurafsky and James Martin (2nd ed., Prentice Hall, 2009). My paper on Naïve Bayes, with Mike Pazzani, is "On the optimality of the simple Bayesian classifier under zero-one loss"* (*Machine Learning*, 1997; expanded journal version of the 1996 conference paper). Judea Pearl's book,* mentioned above, discusses Markov networks along with Bayesian networks. Markov networks in computer vision are the subject of *Markov Random Fields for Vision and Image Processing*,* edited by Andrew Blake, Pushmeet Kohli, and Carsten Rother (MIT Press, 2011). Markov networks that maximize conditional likelihood were introduced in "Conditional random fields: Probabilistic models for segmenting and labeling sequence data,"* by John Lafferty, Andrew McCallum, and Fernando Pereira (*International Conference on Machine Learning*, 2001).

The history of attempts to combine probability and logic is surveyed in a 2003 special issue* of the *Journal of Applied Logic* devoted to the subject, edited by Jon Williamson and Dov Gabbay. "From knowledge bases to decision models,"* by Michael Wellman, John Breese, and Robert Goldman (*Knowledge Engineering Review*, 1992), discusses some of the early AI approaches to the problem.

Chapter Seven

Frank Abagnale details his exploits in his autobiography, *Catch Me If You Can*, cowritten with Stan Redding (Grosset & Dunlap, 1980). The original technical report on the nearest-neighbor algorithm by Evelyn Fix and Joe Hodges is "Discriminatory analysis: Nonparametric discrimination: Consistency properties"* (USAF School of Aviation Medicine, 1951). *Nearest Neighbor (NN) Norms*,* edited by Belur Dasarathy (IEEE Computer Society Press, 1991), collects many of the key papers in this area. Locally linear regression is surveyed in "Locally weighted learning,"* by Chris Atkeson, Andrew Moore, and Stefan Schaal (*Artificial Intelligence Review*, 1997). The first collaborative filtering system based on nearest neighbors is described in "GroupLens: An open architecture for collaborative filtering of netnews,"* by Paul Resnick et al. (*Proceedings of the 1994 ACM Conference on Computer-Supported Cooperative Work*, 1994). Amazon's collaborative filtering algorithm is described in "Amazon.com recommendations: Item-to-item collaborative filtering,"* by Greg Linden, Brent Smith, and Jeremy York (*IEEE Internet Computing*, 2003). (See Chapter 8's further readings for Netflix's.) Recommender systems' contribution to Amazon and Netflix sales is referenced in, among others, Mayer-Schönberger and Cukier's *Big Data* and Siegel's *Predictive Analytics* (cited earlier). The 1967 paper by Tom Cover and Peter Hart on nearest-neighbor's error rate is "Nearest neighbor pattern classification"* (*IEEE Transactions on Information Theory*).

The curse of dimensionality is discussed in Section 2.5 of *The Elements of Statistical Learning*,* by Trevor Hastie, Rob Tibshirani, and Jerry Friedman (2nd ed., Springer, 2009). "Wrappers for feature subset selection,"* by Ron Kohavi and George John (*Artificial Intelligence*, 1997), compares attribute selection methods. "Similarity metric learning for a variable-kernel classifier,"* by David Lowe (*Neural Computation*, 1995), is an example of a feature weighting algorithm.

"Support vector machines and kernel methods: The new generation of learning machines,"* by Nello Cristianini and Bernhard Schölkopf (*AI Magazine*, 2002), is a mostly nonmathematical introduction to SVMs. The paper that started the SVM revolution was "A training algorithm for optimal margin classifiers,"* by Bernhard Boser, Isabel Guyon, and Vladimir Vapnik (*Proceedings of the Fifth Annual Workshop on Computational Learning Theory*, 1992). The first paper applying SVMs to text classification was "Text categorization with support vector machines,"* by Thorsten Joachims (*Proceedings of the Tenth European Conference on Machine Learning*, 1998). Chapter 5 of *An Introduction to Support Vector Machines*,* by Nello Cristianini and John Shawe-Taylor (Cambridge University Press, 2000), is a brief introduction to constrained optimization in the context of SVMs.

Case-Based Reasoning,* by Janet Kolodner (Morgan Kaufmann, 1993), is a textbook on the subject. "Using case-based retrieval for customer technical support,"* by Evangelos Simoudis (*IEEE Expert*, 1992), explains its application to help desks. IPsoft's Eliza is described in "Rise of the software machines" (*Economist*, 2013) and on the company's website. Kevin Ashley explores case-based legal reasoning in *Modeling Legal Arguments** (MIT Press, 1991). David Cope summarizes his approach to automated music composition in "Recombinant music: Using the computer to explore musical style" (*IEEE Computer*, 1991). Dedre Gentner proposed structure mapping in "Structure mapping: A theoretical framework for analogy"* (*Cognitive Science*, 1983). "The man who would teach machines to think," by James Somers (*Atlantic*, 2013), discusses Douglas Hofstadter's views on AI.

The RISE algorithm is described in my paper "Unifying instance-based and rule-based induction"* (*Machine Learning*, 1996).

Chapter Eight

The Scientist in the Crib, by Alison Gopnik, Andy Meltzoff, and Pat Kuhl (Harper, 1999), summarizes psychologists' discoveries about how babies and young children learn.

The *k*-means algorithm was originally proposed by Stuart Lloyd at Bell Labs in 1957, in a technical report entitled "Least squares quantization in PCM"* (which later appeared as a paper in the *IEEE Transactions on Information Theory* in 1982). The original paper on the EM algorithm is "Maximum likelihood from incomplete data via the EM algorithm,"* by Arthur Dempster, Nan Laird, and Donald Rubin (*Journal of the Royal Statistical Society B*, 1977). Hierarchical clustering and other methods are described in *Finding Groups in Data: An Introduction to Cluster Analysis*,* by Leonard Kaufman and Peter Rousseeuw (Wiley, 1990).

Principal-component analysis is one of the oldest techniques in machine learning and statistics, having been first proposed by Karl Pearson in 1901 in the paper "On lines and planes of closest fit to systems of points in space"* (*Philosophical Magazine*). The type of dimensionality reduction used to grade SAT essays was introduced by Scott Deerwester et al. in the paper "Indexing by latent semantic analysis"* (*Journal of the American Society for Information Science*, 1990). Yehuda Koren, Robert Bell, and Chris Volinsky explain how Netflix-style collaborative filtering works in "Matrix factorization techniques for recommender systems"* (*IEEE Computer*, 2009). The Isomap algorithm was introduced in "A global geometric framework for nonlinear dimensionality reduction,"* by Josh Tenenbaum, Vin de Silva, and John Langford (*Science*, 2000).

Reinforcement Learning: An Introduction,* by Rich Sutton and Andy Barto (MIT Press, 1998), is the standard textbook on the subject. *Universal Artificial Intelligence*,* by Marcus Hutter (Springer, 2005), is an attempt at a general theory of reinforcement learning. Arthur Samuel's pioneering research on learning to play checkers is described in his paper "Some studies in machine learning using the game of checkers"* (*IBM Journal of Research and Development*, 1959). This paper also marks one of the earliest appearances in print of the term *machine learning*. Chris Watkins's formulation of the reinforcement learning problem appeared in his PhD thesis *Learning from Delayed Rewards** (Cambridge University, 1989). DeepMind's reinforcement learner for video games is described in "Human-level control through deep reinforcement learning,"* by Volodymyr Mnih et al. (*Nature*, 2015).

Paul Rosenbloom retells the development of chunking in "A cognitive odyssey: From the power law of practice to a general learning mechanism and beyond" (*Tutorials in Quantitative Methods for Psychology*, 2006). A/B testing and other online experimentation techniques are explained in "Practical guide to controlled experiments on the Web: Listen to your customers not to the HiPPO,"* by Ron Kohavi, Randal Henne, and Dan Sommerfield (*Proceedings of the Thirteenth International Conference on Knowledge Discovery and Data Mining*, 2007). Uplift modeling, a multidimensional generalization of A/B testing, is the subject of Chapter 7 of Eric Siegel's *Predictive Analytics* (Wiley, 2013).

Introduction to Statistical Relational Learning,* edited by Lise Getoor and Ben Taskar (MIT Press, 2007), surveys the main approaches in this area. My work with Matt Richardson on modeling word of mouth is summarized in "Mining social networks for viral marketing" (*IEEE Intelligent Systems*, 2005).

Chapter Nine

Model Ensembles: Foundations and Algorithms,* by Zhi-Hua Zhou (Chapman and Hall, 2012), is an introduction to metalearning. The original paper on stacking is "Stacked generalization,"* by David Wolpert (*Neural Networks*, 1992). Leo Breiman introduced bagging in "Bagging predictors"* (*Machine Learning*, 1996) and random forests in "Random forests"* (*Machine Learning*, 2001). Boosting is described in "Experiments with a new boosting algorithm," by Yoav Freund and Rob Schapire (*Proceedings of the Thirteenth International Conference on Machine Learning*, 1996).

"I, Algorithm," by Anil Ananthaswamy (*New Scientist*, 2011), chronicles the road to combining logic and probability in AI. *Markov Logic: An Interface Layer for Artificial Intelligence*,* which I cowrote with Daniel Lowd (Morgan & Claypool, 2009), is an introduction to Markov logic networks. The Alchemy website, http://alchemy.cs.washington.edu, also includes tutorials, videos, MLNs, data sets, publications, pointers to other systems, and so on. An MLN for robot mapping is described in "Hybrid Markov logic networks,"* by Jue Wang and Pedro Domingos (*Proceedings of the Twenty-Third AAAI Conference on Artificial Intelligence*, 2008). Thomas Dietterich and Xinlong Bao describe the use of MLNs in DARPA's PAL project in "Integrating multiple learning components through Markov logic"* (*Proceedings of the Twenty-Third AAAI Conference on Artificial Intelligence*, 2008). "Extracting semantic networks from text via relational clustering,"* by Stanley Kok and Pedro Domingos (*Proceedings of the Nineteenth European Conference on Machine Learning*, 2008), describes how we used MLNs to learn a semantic network from the Web.

Efficient MLNs with hierarchical class and part structure are described in "Learning and inference in tractable probabilistic knowledge bases,"* by Mathias Niepert and Pedro Domingos (*Proceedings of the Thirty-First Conference on Uncertainty in Artificial Intelligence*, 2015). Google's approach to parallel gradient descent is described in "Large-scale distributed deep networks,"* by Jeff Dean et al. (*Advances in Neural Information Processing Systems 25*, 2012). "A general framework for mining massive data streams,"* by Pedro Domingos and Geoff Hulten (*Journal of Computational and Graphical Statistics*, 2003), summarizes our sampling-based method for learning from open-ended data streams. The FuturICT project is the subject of "The machine that would predict the future," by David Weinberger (*Scientific American*, 2011).

"Cancer: The march on malignancy" (*Nature* supplement, 2014) surveys the current state of the war on cancer. "Using patient data for personalized cancer treatments," by Chris Edwards (*Communications of the ACM*, 2014), describes the early stages of what could grow into CanceRx. "Simulating a living cell," by Markus Covert (*Scientific American*, 2014), explains how his group built a computer model of a whole infectious bacterium. "Breakthrough Technologies 2015: Internet of DNA," by Antonio Regalado (*MIT Technology Review*, 2015), reports on the work of the Global Alliance for Genomics and Health. Cancer Commons is described in "Cancer: A Computational Disease that AI Can Cure," by Jay Tenenbaum and Jeff Shrager (*AI Magazine*, 2011).

Chapter Ten

"Love, actuarially," by Kevin Poulsen (*Wired*, 2014), tells the story of how one man used machine learning to find love on the OkCupid dating site. *Dataclysm*, by Christian Rudder (Crown, 2014), mines OkCupid's data for sundry insights. *Total Recall*, by Gordon Moore and Jim Gemmell (Dutton, 2009), explores the implications of digitally recording everything we do. *The Naked Future*, by Patrick Tucker (Current, 2014), surveys the use and abuse of data for prediction in our world. Craig Mundie argues for a balanced approach to data collection and use in "Privacy pragmatism" (*Foreign Affairs*, 2014). *The Second Machine Age*, by Erik Brynjolfsson and Andrew McAfee (Norton, 2014), discusses how progress in AI will shape the future of work and the economy. "World War R," by Chris Baraniuk (*New Scientist*, 2014) reports on the debate surrounding the use of robots in battle. "Transcending complacency on superintelligent machines," by Stephen Hawking et al. (*Huffington Post*, 2014), argues that now is the time to worry about AI's risks. Nick Bostrom's *Superintelligence* (Oxford University Press, 2014) considers those dangers and what to do about them.

A Brief History of Life, by Richard Hawking (Random Penguin, 1982), summarizes the quantum leaps of evolution in the eons BC. (Before Computers. Just kidding.) *The Singularity Is Near*, by Ray Kurzweil (Penguin, 2005), is your guide to the transhuman future. Joel Garreau considers three different scenarios for how human-directed evolution will unfold in *Radical Evolution* (Broadway Books, 2005). In *What Technology Wants* (Penguin, 2010), Kevin Kelly argues that technology is the continuation of evolution by other means. *Darwin Among the Machines*, by George Dyson (Basic Books, 1997), chronicles the evolution of technology and speculates on where it will lead. Craig Venter explains how his team synthesized a living cell in *Life at the Speed of Light* (Viking, 2013).

Index

PEDRO DOMINGOS is a professor of computer science at the University of Washington. He is a winner of the SIGKDD Innovation Award, the highest honor in data science. A fellow of the Association for the Advancement of Artificial Intelligence, he lives near Seattle.